Advances in Sustainability Science and Technology

Series Editors

The book series aims at bringing together valuable and novel scientific contributions that address the critical issues of renewable energy, sustainable building, sustainable manufacturing, and other sustainability science and technology topics that have an impact in this diverse and fast-changing research community in academia and industry.

The areas to be covered are

- Climate change and mitigation, atmospheric carbon reduction, global warming
- Sustainability science, sustainability technologies
- Sustainable building technologies
- Intelligent buildings
- Sustainable energy generation
- Combined heat and power and district heating systems
- Control and optimization of renewable energy systems
- Smart grids and micro grids, local energy markets
- Smart cities, smart buildings, smart districts, smart countryside
- Energy and environmental assessment in buildings and cities
- Sustainable design, innovation and services
- Sustainable manufacturing processes and technology
- Sustainable manufacturing systems and enterprises
- Decision support for sustainability
- Micro/nanomachining, microelectromechanical machines (MEMS)
- Sustainable transport, smart vehicles and smart roads
- Information technology and artificial intelligence applied to sustainability
- Big data and data analytics applied to sustainability
- Sustainable food production, sustainable horticulture and agriculture
- Sustainability of air, water and other natural resources
- Sustainability policy, shaping the future, the triple bottom line, the circular economy

High quality content is an essential feature for all book proposals accepted for the series. It is expected that editors of all accepted volumes will ensure that contributions are subjected to an appropriate level of reviewing process and adhere to KES quality principles.

The series will include monographs, edited volumes, and selected proceedings.

V. Bindhu · João Manuel R. S. Tavares · Stefan Talu
Editors

Proceedings of Sixth International Conference on Inventive Material Science Applications

ICIMA 2023

Springer

Editors
V. Bindhu
PPG Institute of Technology
Coimbatore, Tamil Nadu, India

João Manuel R. S. Tavares
Departamento de Engenharia Mecânica
Universidade do Porto
Porto, Portugal

Stefan Talu
The Technical University of Cluj-Napoca
Cluj-Napoca, Romania

ISSN 2662-6829 ISSN 2662-6837 (electronic)
Advances in Sustainability Science and Technology
ISBN 978-981-99-4188-9 ISBN 978-981-99-4189-6 (eBook)
https://doi.org/10.1007/978-981-99-4189-6

This Springer imprint is published by the registered company Springer Nature Singapore Pte Ltd.
The registered company address is: 152 Beach Road, #21-01/04 Gateway East, Singapore 189721, Singapore

The conference is dedicated to the outstanding reviewers, authors, editors, and organizers of the conference to commemorate their inevitable contributions to advanced materials science and related fields. Without their participation, it would have been impossible to hold the 6th ICIMA 2023 successfully and ensure high quality papers were published in the conference proceedings.

Preface

We are very pleased to introduce the Proceedings of the 6th International Conference on Inventive Material Science Applications (ICIMA 2023). The 6th ICIMA event was held at PPG Institute of Technology, from 11th to 12th May 2023.

One of the significant and valuable dimensions of this 6th conference edition is the way to researchers, academicians, and engineers in the field from different countries come to one place and initiate discussion on the relevant issues, challenges, opportunities, and research findings. The primary focus of ICIMA 2023 is to provide an excellent platform for the conference participants to share and exchange novel and innovative ideas of original research, and to build international association. The main intent of this conference is to create a smart and advanced research landscape for the areas of inventive material science.

Out of 46 submissions received from different parts of the world, only 13 submissions were accepted as full papers for publication and presentation in ICIMA 2023. These papers provide a brief illustration of current research on relevant topics, covering sustainable strategy, micro/nanomaterials, biomaterials, hybrid electronic materials, innovative electronic materials processing, computational material science, material characterization, fabrication, and synthesis technologies.

The success of the conference is due to the collective efforts of all the reviewers and advisory/review board members. We would like to express and record our gratitude and appreciation to the authors for their contributions. Many thanks to the reviewers, who helped us to maintain high-quality manuscripts included in the proceedings. We also express our sincere thanks to the keynote member 'Dr. M. V. Reddy, Institute of Research Hydro Quebec, Canada', members of the conference committees and organizing team for their hard work. We wish that all the authors and delegates find ICIMA 2023 proceedings interesting, exciting, and inspiring.

Dr. V. Bindhu
Professor and Head, ECE
PPG Institute of Technology
Coimbatore, India

Dr. João Manuel R. S. Tavares
Professor
Universidade do Porto (FEUP)
Porto, Portugal

Dr. Stefan Talu
The Technical University
of Cluj-Napoca
Cluj-Napoca, Romania

Contents

About the Editors

V. Bindhu received B.E. degree in Electronics and Communication Engineering from Bharathiar University, Coimbatore, in 2002, M.E. degree in Applied Electronics from Anna University, Chennai, in 2007, and Ph.D. degree from Anna University, Chennai, in 2014. She has 10 years of teaching experience and 5 years of research experience. Currently, she is a professor at PPG Institute of Technology, Coimbatore. Her areas of interest include signal processing and VLSI design.

João Manuel R. S. Tavares has a B.Sc. degree in Mechanical Engineering, M.Sc. and Ph.D. degrees in Electrical and Computer Engineering, and Habilitation in Mechanical Engineering. He is the Editor-in-Chief of the Computer Methods in Biomechanics and Biomedical Engineering: Imaging & Visualization and Computer Methods in Biomechanics and Biomedical Engineering journals and the editor of the book series Lecture Notes in Computational Vision and Biomechanics.

Stefan Talu the holder of Ph.D. degree in Engineering, is currently teaching Descriptive Geometry and Engineering Graphics at the Technical University of Cluj-Napoca, Cluj-Napoca, Romania. He obtained his Ph.D. degree in Technical Sciences, in 1998, from the Technical University of Cluj-Napoca, Romania. He has published 19 books and 380 scientific papers (200 papers as a single author or first author), in journals, at symposia, and scientific national and international conferences, in country or abroad, among them 99 papers in Clarivate Analytics indexed journals, with impact factor and relative influence score, 146 papers in international databases, 40 papers at national conferences and 95 papers at international conferences, in collaboration with 210 researchers from 27 foreign countries. He is an expert in the following fields: mechanical and tribological characterizations of macro-micro-nanostructures; topographical and morphological characterizations of three-dimensional surfaces; computer-aided engineering and designing industrial processes, and fractal and multifractal geometry applied in technique.

Chapter 1
Experimental Investigation of Tribological Characterization of Bio-Nanolubricants

Ravikiran, S. L. Aravind, B. G. Chetan, Afnaan Ali Khan, Prashanth, H. Poornananda, K. S. Sathvik, and K. Shriharsha

Abstract The growing environmental concerns in recent years have led to the acceptance of vegetable oil bio-based lubricants as an acceptable replacement for petroleum-based lubricants. In this article, we analyze the tribological properties of three different bio-nano-lubricants with ZrO2, TiO2, and ZnO as additives, and explain how the nanoparticles diffuse into mixed vegetable oils. Lubricant samples are prepared by mixing two different vegetable oils, such as castor oil and sunflower oil, in different ratios, then dispersing the nanoparticles in the mixed oil and sonicating to ensure homogeneity of the mixture using a magnetic stirrer. The viscosity is determined using a Redwood viscometer. Measurement of the flash point and ignition point is done with the Pensky–Martin Closed Cup Tester. All friction and antiwear properties of lubricating fluids are measured using a four-ball tester according to ASTM D 4172–92. Results are compared with several mineral oils. The use of nanoparticles significantly improves friction and wear properties compared to other types of mineral oils. The results of this study also showed superior levels of other attributes, including viscosity, when compared to mineral oil-based lubricants.

Keywords Nanolubricants · Nanoparticles · Viscosity · Flash point and Fire point · Four-ball tester · Friction wear

---

Ravikiran (✉) · S. L. Aravind · B. G. Chetan · A. Ali Khan · Prashanth · H. Poornananda · K. S. Sathvik · K. Shriharsha
Department of Mechanical Engineering, The National Institute of Engineering, Mysore, Karnataka, India
e-mail: nayak.ravikiran@gmail.com

S. L. Aravind
e-mail: aravindsl@nie.ac.in

V. Bindhu et al. (eds.), *Proceedings of Sixth International Conference on Inventive Material Science Applications*, Advances in Sustainability Science and Technology,
https://doi.org/10.1007/978-981-99-4189-6_1

1 Introduction

Lubricants are essential products that are made from mineral oils. These oils are utilized for a variety of things, including the efficient operation of all machinery and their tools [1]. Since long before the emergence of modern technology, humans have sought a substance to minimize friction between surfaces that come into contact. For a machine to operate effectively and for a prolonged period, and friction should be minimized as it contributes to heat generation, component wear, and other problems. Lubricants are available in a variety of forms, including solid, liquid, dry film, and coating [2, 3].

Often these lubricants used worldwide end up harming the environment because they're poisonous and hazardous to soil and aquatic species. Global climate change is caused by a rise in greenhouse gas concentrations in the atmosphere as a result of the increasing use of mineral oil-based lubricants. The usage of bio-lubricants is becoming more popular in the lubricating industry around the world in an attempt to reduce all current environmental concerns [4, 5].

Vegetable oils possess a great degradation rate, low temperature stability, high viscosity index, and reduced toxicity compared to mineral resources. There are two possible ways to use vegetable oils: one is to simply use the best vegetable oils alone, without mixing; the other is to blend the vegetable oils in a certain proportion with the industrial lubricants [6, 7]. In comparison to other mineral-based oils, vegetable oils have a greater lubricating performance and have the lowest frictional coefficient [8].

Various studies have been focusing on improving the machining capabilities of vegetable oil by including nanoparticles [9, 10]. Vegetable oils perform better when nanoparticles are added, because nanoparticles minimize friction coefficient and cutting temperature, in addition, results showed that the viscosity, friction coefficient, and wear rate were all improved by the inclusion of nanoparticles [11]. The effect of incorporating ZnO, CuO, and ZrO2 nanoparticles into oils was discovered to reduce friction and wear due to the nanoparticles serving as load-bearings.

This paper aims to study the various enhancing lubricating properties by blending vegetable oils with mineral oxide nanoparticles [12, 13]. It is important to mention here that the investigated present work focuses on an enhanced high-grade lubricating oil such as SAE 15W40 engine oil [14, 15]. Therefore, the main aim present work is to study the effect of adding 0.2–1% percentage variation of TiO2 into castor oil (70–75%) and sunflower oil (24–30%) in comparison with the results of 0.25–1% percentage variation of ZrO_2 nanoparticles into the blend of castor oil (70–90%) and sunflower oil (10–30%) on the properties such as viscosity, coefficient of viscosity, flash point, fire point of nanolubricants [16].

2 Methodology

2.1 Sample Preparation

Castor oil and sunflower oil are used as basic oils, and TiO_2 and ZrO_2 are added as nanoparticles in various combinations to create nanolubricants. The mixture was heated to 60 °C and agitated constantly for 40 min at 450–500 rpm using a magnetic stirrer, as shown in Fig. 1. The concoction was then agitated constantly, and air quenched to bring it to room temperature. The different combinations of basic oil and nanoadditives are illustrated in Tables 1 and 2.

Fig. 1 Magnetic stirrer

Table 1 Castor oil + Sunflower oil + TiO_2→(100 gm)

Sl. no	Castor oil (gm)	Sunflower oil (gm)	Tio_2 (gm)
1	70	30	0.25
2	70	30	0.5
3	75	30	1

Table 2 Castor oil + Sunflower oil + ZrO_2→(100 gm)

Sl no	Castor oil (gm)	Sunflower oil (gm)	ZrO_2 (gm)
1	70	30	0.25
2	70	30	0.5
3	75	30	1

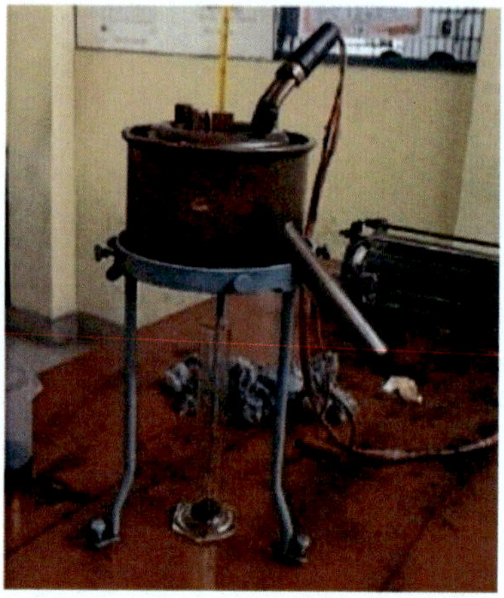

Fig. 2 Redwood viscometer

2.2 Absolute Viscosity Measurement

A viscometer, as shown in Fig. 2, is the Redwood Viscometer, which measures how long it takes an oil sample to flow through a standard opening called an orifice and is then collected in a 50 ml beaker, giving a numerical value for the oil's viscosity in Redwood seconds. The kinematic viscosity of the prepared oil is calculated using the formula [17]

$$\nu = (0.00247t - (0.65/t))$$

The absolute viscosity is calculated using the formula μ = kinematic viscosity * mass density of the oil at corresponding temperature.

2.3 Flash and Fire Points

The Pensky–Martens closed-cup flash point is an instrument for determining the flash point of flammable liquids, as shown in Fig. 3. In this oil, sample is poured into a cup. The meter is placed in oil, and the oil is heated. The test flame is passed over the cup at every 2 °C increase in sample temperature. When the sample vapors ignite momentarily in air [18], the flash point temperature is reached. When the sample vapors sustain combustion for at least five seconds, the fire point temperature is reached.

Fig. 3 Pensky–Martin's closed-cup tester

Table 3 Test parameters

Parameters	Condition
Load	40 kg
Temperature	75 °C
Speed	1200 rpm
Time	60 min

2.4 Four-Ball Test

The test was performed on the following samples according to ASTM D 4172–92.The balls used comprised a diameter of 12.7 mm, a mirror finish, and an HRC of 64–66. These balls were composed of chromium steel manufactured by SKF. With the aid of an optical microscope and a four-ball tester, the value of the coefficient of friction and wear scar diameter were measured. Test parameters are listed in Table 3.

3 Results and Discussions

The outcomes of testing samples under various circumstances are compared to those of mineral oils and bio-lubricants.

3.1 Viscosity Measurement

From the tables and figures (Figs. 5 and 6), it can be observed that the viscosity decreases with the rise in temperature as usual; therefore, the viscosity of lubricating

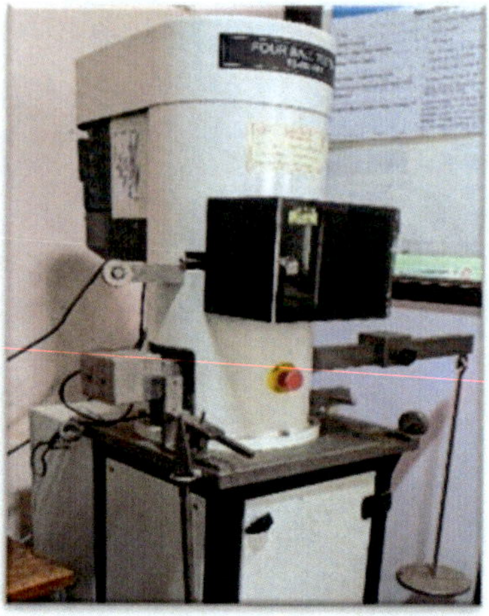

Fig. 4 Four-ball test apparatus

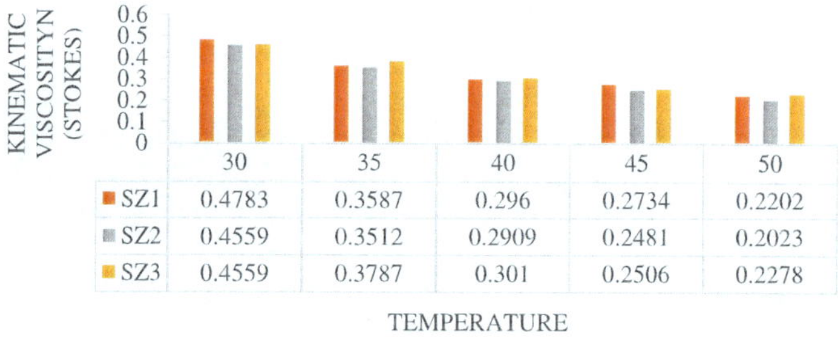

Fig. 5 Kinematic viscosity versus temperature graph for Sunflower oil + ZnO NP

oil also varies. Tables 4 and 5 show that the viscosity of ZnO NP dispersed sunflower oil was found to be higher than the TiO2 NP dispersed castor and sunflower blended oil and an increase in temperature enhances the movements of the molecules and reduces intermolecular forces so the layers of liquid easily pass over one on another and thus contribute to the reduction of viscosity.

3.2 Flash and Fire Points

Tables 6 and 7 show the flash point and fire point parameters.

From Fig. 7, it is found that SZ3 shows the high flash and fire point followed by SZ2 and SZ1. From Fig. 8, it is found that CST1 shows high flash and fire points followed by CST3 and CST2. Combining both results, TiO2 NP dispersed caster

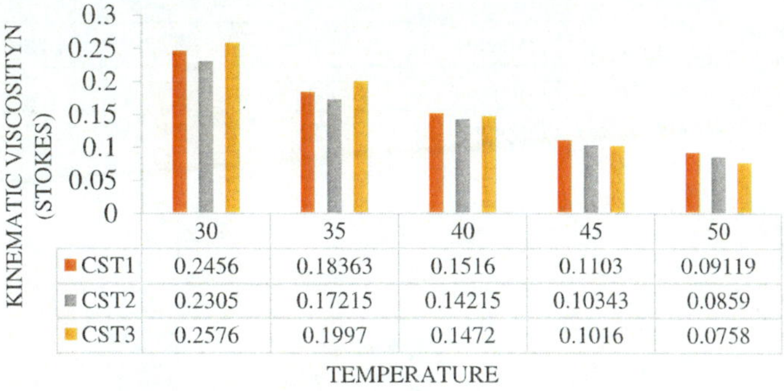

Fig. 6 Kinematic viscosity versus temperature graph for Castor oil+ Sunflower oil + TiO2 NP

Table 4 Sunflower oil + ZnO nanoparticles

Temperature (°C)	SZ1		SZ2		SZ3	
	Kinematic viscosity	Absolute viscosity	Kinematic viscosity	Absolute viscosity	Kinematic viscosity	Absolute viscosity
	ν (stokes)	μ (poise)	ν (stokes)	μ (poise)	ν (stokes)	μ (poise)
30	0.4783	0.4194	0.4559	0.4359	0.4559	0.4559
35	0.3587	0.3145	0.3512	0.3358	0.3787	0.3512
40	0.296	0.2595	0.2909	0.2782	0.301	0.2909
45	0.2734	0.2397	0.2481	0.2372	0.2506	0.2481
50	0.2202	0.1931	0.2023	0.1934	0.2278	0.2023

Table 5 Castor oil + sunflower oil + TiO2 NP

Temperature (°C)	CST1		CST2		CST3	
	Kinematic viscosity	Absolute viscosity	Kinematic viscosity	Absolute viscosity	Kinematic viscosity	Absolute viscosity
	ν (stokes)	μ (poise)	ν (stokes)	μ (poise)	ν (stokes)	μ (poise)
30	0.2456	0.2305	0.27151	0.2576	0.2365	0.2206
35	0.18363	0.17215	0.20982	0.1997	0.1909	0.1779
40	0.1516	0.14215	0.15512	0.1472	0.1466	0.1366
45	0.1103	0.10343	0.10708	0.1016	0.0995	0.0927
50	0.09119	0.0859	0.07895	0.0758	0.07396	0.0689

Table 6 Sunflower oil + ZnO NP

Parameter	CST1	CST2	CST3
Flash point	220	230	246
Fire point	232	240	254

Table 7 Castor oil + Sunflower oil + TiO2 NP

Parameter	SZ1	SZ2	SZ3
Flash point	220	230	246
Fire point	232	240	254

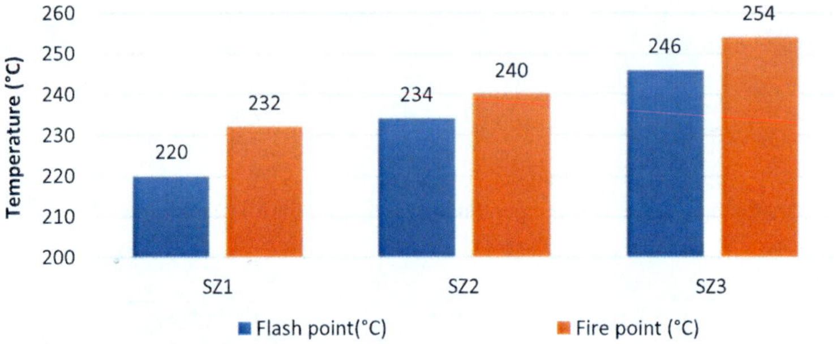

Fig. 7 Flash and fire points of sunflower oil samples

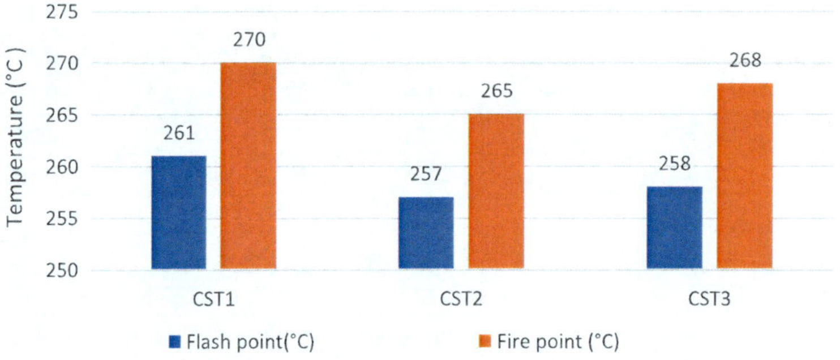

Fig. 8 Flash and fire points of castor oil samples

sunflower blend (CST) shows a higher value of flash and fire points compared to ZnO NP dispersed sunflower oil (SZ).

3.3 Tribological Properties

One of the important factors to consider when examining the tribological properties of the tested nanolubricants is the coefficient of friction (COF). For all of the examined lubricants with the nanoparticles, decreased friction coefficients were observed as well in this investigation. For the first 10 min of the wear process, the friction coefficients increased; following which friction reduced. The data on the friction coefficient shown in Fig. 9 has been collected during the four-ball test. However, the

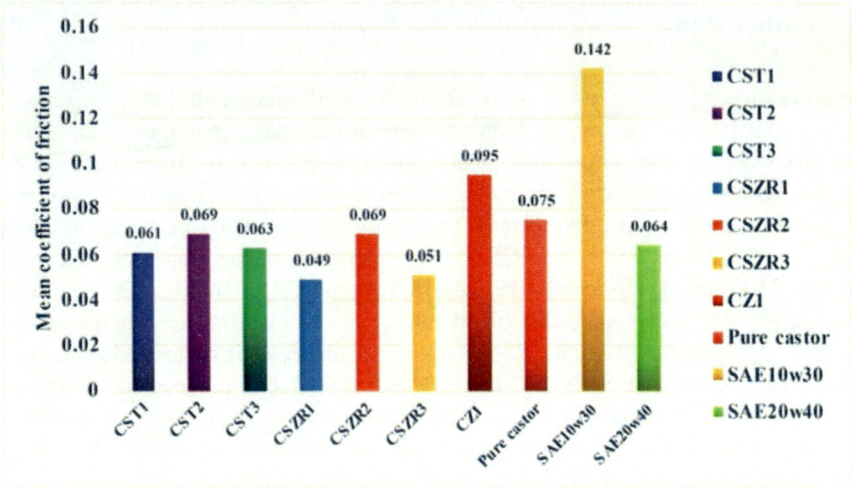

Fig. 9 Mean coefficient of friction of sunflower, castor oil, and SAE grade lubricants

peak values of the friction coefficient in the tests with ZnO were lower than those in the tests with other nanoparticles during the first 10 min of testing. The presence of the nanoparticles could reduce the immediate contact between the ball and disc and make up for any nanodefects on the first contact surfaces.

From the mean friction coefficient values shown in Fig. 9, CSZR1 and CSZR3 exhibit the lowest friction coefficient value of 0.049 and 0.051. When less %wt. ZrO_2 dispersed castor sunflower blended nanolubricant is compared to pure castor oil, the coefficient of friction is reduced approximately by 40%, but $CSZr_2$ exhibits a little high coefficient of friction compared to CSZR1 and CSZR3. In the case of TiO_2 nanoparticles dispersed castor and sunflower blend, all three CST1, CST2, and CST3 samples exhibit a similar coefficient of friction 0.061, 0.069, and 0.063, respectively. When the mean COF of CSZR1 and CSZR3 is compared with the mean coefficient of friction of synthetic lubricants like SAE10W30 and SAE20W40, even TiO_2 dispersed castor sunflower oil blended lubricants show a similar friction coefficient with SAE10W30 and SAE20W40 but little high friction coefficient than ZrO_2 nanoparticles dispersed castor sunflower oil blended lubricants. The ZnO nanoparticles are dispersed ssn flower oil alone shows a high coefficient of friction compared to all other sample tests. The variation of friction coefficient in Fig. 9 is considered to be independent of the working conditions and temperature.

4 Conclusion

In the lubricating market, most lubricants are made of mineral oil. Lubricants made from mineral oil have many disadvantages after use. It causes environmental pollution like soil pollution, air pollution, water pollution, etc. The quality and properties of vegetable oil blends by adding nanoparticles as additives were evaluated through this study using different parameters. This study shows that when the mean coefficient of friction of CSZr1 and $CSZr_3$ is compared with the mean coefficient of friction of synthetic lubricants like SAE10W30 and SAE20W40, ZrO_2 nanoparticles dispersed castor sunflower oil blended lubricants shows less coefficient of friction. Lubricants containing ZrO_2 nanoparticles dispersed in castor and sunflower oils perform better than SAE10W30 and SAE20W40 due to reduced coefficient of friction and lower value of wear scar diameter. Followed by TiO2 nanoparticles dispersed castor sunflower blended lubricants which show a similar coefficient of friction with SAE10W30 and SAE20W40 and a better coefficient of friction than pure castor oil. Also, wear scar diameter for the 1% wt. ZrO_2 nanoparticles dispersed castor sunflower blended oil has a lower value at 70:30 ratio. As a result, it is possible that a mixture of castor oil, sunflower oil, and ZrO_2 nanoparticles will work as an anti-wear lubricant. According to analysis, castor oil blends with nanoparticles scattered in them may one day serve as a partial replacement for bio-lubricants. This is due to the fact that blends have no adverse impacts on wear phenomena or lubrication functioning.

References

1. Chandrakar JK, Suhane A (2014) The prospects of vegetable based oils as metal working fluids in manufacturing application–a review. Int J Eng Res Technol (IJERT) 03(05)
2. Karmakar G, Ghosh P, Sharma BK (2017) Chemically modifying vegetable oils to prepare green lubricants. Lubricants 5(4):44. https://doi.org/10.3390/lubricants5040044
3. Chinchkar DS, Satpute ST, Kumbhar NR (2012) Castor oil as green lubricant: a review. Int J Eng Res Technol (IJERT) 01(05)
4. Talkit KM, Mahajan DT, VH Masand (2015) Analytical studies on lubrication properties of different vegetable oils blends at different temperatures
5. Kannan I, Karthikeyan S, Ravikumar K (2020) Lubrication performance of castor oil blended with other vegetable oils in grinding of inconel 625. IOP Conf Ser Mater Sci Eng 923:012027. https://doi.org/10.1088/1757-899X/923/1/012027
6. Husain A, Mehdi M, Ali A, Adeel M, Hassan M, Ani F (2018) Investigation the tribological characteristics of castor oil with mineral oil blends. J Eng Appl Sci 37:1–6
7. Jhodkar D, Gupta K Effect of CuO nanoparticles-based vegetable oil on the machining performance of WC tool insert while turning EN 8 steel
8. Cortes V, Sanchez K, Gonzalez R, Alcoutlabi M, Ortega JA (2020) The performance of SiO2 and TiO2 nanoparticles as lubricant additives in sunflower oil. Lubricants 8(1):10. https://doi.org/10.3390/lubricants8010010
9. Ta T-N, Chern S-Y, Horng J-H (2021) Tribological behavior of ionic liquid with nanoparticles. Materials 14:6318. https://doi.org/10.3390/ma14216318

10. Mallick P, Sahu S (2012) Structure, microstructure and optical absorption analysis of CuO nanoparticles synthesized by sol-gel route. Nanosci Nanotechnol 2:71–74. https://doi.org/10.5923/j.nn.20120203.05
11. Bekru A, Zelekew OA, Andoshe DM, Sabir FK, Eswaramoorthy R (2021) Microwave-assisted synthesis of CuO nanoparticles using cordia africana Lam. Leaf extract for 4-nitrophenol reduction. J Nanotechnol 1–12. https://doi.org/10.1155/2021/5581621
12. Silva N, Ramírez S, Díaz I, García A, Hassan N (2019) Easy, quick, and reproducible sonochemical synthesis of CuO nanoparticles
13. Hasnidawani JN, Azlina HN, Norita H, Bonnia NN, Ratim S, Ali SE (2016) Synthesis of ZnO nanostructures
14. Al Abdullaha K, Awada S, Zaraketb J, Salameb C (2017) Synthesis of ZnO nano powders by using sol-gel and studying their structural and electrical properties at different temperature
15. Savi BM, Rodrigues L, Bernardin AM (2011) Synthesis of Zno nanoparticles by sol-gel processing
16. Shahabuddin M, Masjuki HH, Kalam MA (2013) Experimental investigation into tribological characteristics of bio-lubricant formulated from jatropha oil. Proc Eng 56:597–606. https://doi.org/10.1016/j.proeng.2013.03.165
17. Li X, Murashima M, Umehara N (2018) Effect of nanoparticles as lubricant additives on friction and wear behavior of tetrahedral amorphous carbon (ta-C) coating. J Tribologi 16:15–29
18. Syahrullail S, et al (2013) Performance of vegetable oil using sol-gel method. Proc Chem 19:211–216. https://doi.org/10.1016/j.proche.2016.03.095 as Lubricant in extreme pressure condition. Proc Eng 68(2013):172–177

Chapter 2
Effect of Eu^{3+} on the Luminescence and Photocatalytic Properties of ZnS Nanoparticles

Lallianmawii and N. Mohondas Singh

Abstract Eu^{3+} doped ZnS was prepared using a reflux method. Using different techniques such as XRD, SEM, TEM, FTIR, PL, and UV, the structural, morphological, and optical properties were examined. Photoluminescence studies show the most intense electric dipole transition $^5D_0 \rightarrow {}^7F_2$, which is particularly sensitive to the chemical bond in the vicinity of an ion of europium (III) when excited at 350 nm. The CIE studies show all the prepared samples fall in the white regions, which have potential applications in optical devices, lightning outdoors, etc. The degradation percentage of methyl red was found to be 90% and follows first-order kinetics. Additionally, the Eu^{3+} plays a crucial role in the production of free radicals. As a result, the catalyst has enhanced degradation activity, high stability, and can be utilized for the degradation of toxic dyes present in the environment.

Keywords Photoluminescence · Degradation · Methyl red · Catalyst · Dyes · Environment

1 Introduction

Environmental contaminants are often regarded as one of the main problems that have a negative impact on human health. Water evaporation, surface water contamination, and drinking water shortages have been identified as major global issues in recent years [1]. The pharmaceutical, chemical, and textile industries, which carelessly dump their waste into the environment and water bodies, are the primary sources of various organic pollutants that contaminate water. The ecosystem and aquatic life are severely harmed by these hazardous organic effluents [2]. The integrity of the world's water supply has gotten worse due to the numerous colours and toxic organic pollutants in aquatic environments [3]. The fast growth of the textile industry has also

Lallianmawii · N. M. Singh (✉)
Department of Chemistry, Mizoram University, Tanhril, Mizoram, India
e-mail: nmdas08@rediffmail.com

V. Bindhu et al. (eds.), *Proceedings of Sixth International Conference on Inventive Material Science Applications*, Advances in Sustainability Science and Technology,
https://doi.org/10.1007/978-981-99-4189-6_2

raised concerns about water contamination due to the durability and toxicity of these dyes to both aquatic and terrestrial organisms [4]. When compared to primary effluent treatment methods, which are less successful and also create secondary pollutants, semiconductor photocatalysis is exceptionally effective at completely breaking down organic contaminants [5].

Dyes are frequently employed as colouring agents in a variety of sectors, including cloth, fleece, food, plastics, and cosmetics [6]. The dyes are intricate, water-soluble chemical compounds with low biodegradability that are also potentially carcinogenic and mutagenic. Natural dyes and synthetic/artificial dyes are the two main categories of dyes [7]. Natural dyes are derived from plant materials, including bark, leaves, wood, roots, etc., whereas synthetic dyes are created from substances like chemicals, minerals, and petroleum derivatives and are used in the paper, textile, and cosmetic industries as colouring agents and printing processes [8]. Synthetic dyes are organic molecules that are identified by chromophores in their molecular structures. They are frequently divided into a number of classes based on the chromophore groups they contain, such as azo, anthraquinone, indigoide, phthalocyanine, sulphur, and triphenylmethane derivatives [9, 10]. Organic colours that are highly cytotoxic to mammalian tissues, including azo and fluorescein dyes, are difficult to break down naturally. Based on the sort of change they have, dyes are also divided into two major categories: non-ionic dyes and ionic dyes [11]. The two types of ionic dyes are cationic and anionic, which have positive and negative charges, respectively. Due to their limited biodegradability, organic dyes have produced significant environmental contamination issues. So, it is crucial for environmental protection that dye molecules are removed from industrial effluent. Diverse technologies, such as carbon adsorption, flocculation, ozonization, and activated sludge procedures, are currently available to treat industrial effluent dyes. Nevertheless, these processes are expensive, and it has been discovered that the degradation kinetics are slow and ineffective [12]. Advanced oxidation processes (AOPs), which use photocatalytic reactions to produce superoxide (O_2), reactive oxygen species (ROS), and hydroxyl (OH) radicals, are promising techniques for eliminating dye contaminants [13]. With its efficiency and wide applicability, photocatalysis is an environmentally friendly and green process that has the potential to completely remove toxic pollutants from the environment. It also has a number of advantages, such as the capacity to mineralize hazardous organic pollutants into non-toxic inorganics like carbon dioxide and water and to break down toxins without the use of secondary chemicals [14].

Due to its superiority, environmental friendliness, efficacy, and good stability, photocatalytic degradation of semiconductors serving as photocatalysts is acknowledged as a highly successful method of treating antibiotic residues [15]. ZnS is an n-type semiconductor photocatalyst with a suitable negative redox potential in the conduction band and valence band and a large bandgap of 3.68 eV. It has remarkable chemical stability against oxidation and hydrolysis [16]. Due to their lack of toxicity, ZnS nanoparticles are the preferred material for use in biological and photocatalytic processes. ZnS has recently been thoroughly investigated as an innovative semiconductor in the fields of photocatalysis, sensing, and imaging due to its significant qualities, such as visible-light responsiveness, acceptable band locations, greater

Table 1 Physiochemical properties of methyl red

Sample	Molecular formula	Molecular structure	Density	M_w(g/mol)
Methyl red	$C_{15}H_{15}N_3O_2$	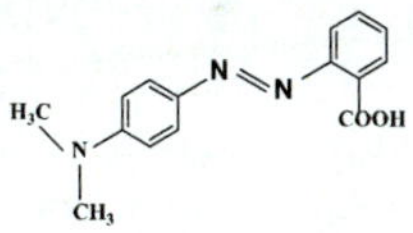	0.839 g/mL–25 °C	269.3 g/mol

chemical and thermal stability, and non-toxic and affordable manufacture [17]. The lower hazardous content, higher surface area, commercial viability, biocompatibility, and chemical inertness of the nanoparticles make the modification with transition metal ion doping significant [18]. Doping is a common and efficient method for changing the energy levels, surface states, and structural, optical, magnetic, and spintronic properties of semiconductors [19]. Owing to the critical role that dopants play in semiconductor devices, researchers have been looking into the properties and prospective applications of these semiconductor nanoparticles that have been purposefully doped. In this study, a reflux approach was used to synthesize Eu^{3+} doped ZnS nanoparticles. The characteristics of ZnS nanoparticles are altered by the addition of Eu^{3+} dopants. Physiochemical properties of Methyl red is shown in Table 1.

2 Experimental Section

2.1 Chemical Used

Europium (III) nitrate hexahydrate ($EuH_{12}N_3O_{15}$), Alfa Aesar purity 99.9%, Zinc acetate dehydrate $(CH_3COO)_2$ Zn + $2H_2O$ Alfa Aesar, purity 98.0%, and Sodium Sulfide H_2Na_2OS purity 98.0% Alfa Aesar was used to synthesize Eu^{3+} doped ZnS.

2.2 Synthesis

Eu^{3+} doped ZnS (2, 4, 6 at.%) was synthesized using the reflux method. Zinc acetate dehydrate and europium (III) nitrate hexahydrate were dissolved in 50 ml of distilled water and stirred for 1 h at room temperature. After stirring at room temperature, sodium sulfide was added drop by drop, and the solution was refluxed at 150 °C for 3 h. The obtained precipitate was ultrasonically treated for 30 min. Finally, the solution was centrifuged at 7000 rpm. After drying, the samples were used for characterization.

2.3 *Characterization*

Rigaku Cu K α Smart Lab was used to obtain XRD data. Image from TEM was taken by FEI Tecnai (G2 F20 X-TWIN). Supa 55 FESEM model with a high resolution was used to capture the FESEM image. Shimadzu IR affinity-1S FT-IR data were collected. Photoluminescence studies were done using a Hitachi F-7000 fluorescence spectrometer. Thermo-Scientific evolution of 220 was used to study UV visible spectrophotometer.

3 Results and Discussions

3.1 *XRD*

Figure 1 displays the XRD patterns of Eu^{3+} doped ZnS nanoparticles. The planes (111), (220), and (311) of the face-centred cubic (FCC) of ZnS were observed as three large peaks corresponding to (JCPDS No.05–0566). The Debye–Scherrer equation is used to determine the particle size [16]:

$$D = K\lambda/\beta cos\theta$$

Fig. 1 X-ray diffraction of Eu^{3+} doped ZnS 2, 4, 6 at.%

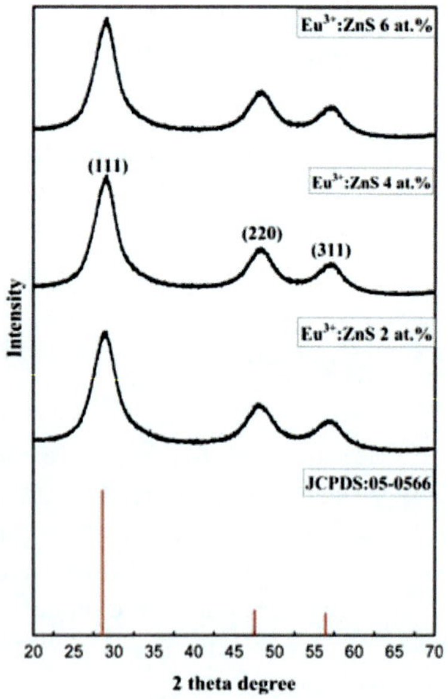

Table 2 Calculation of crystal size, cell volume, and lattice parameters from XRD

Sample	Lattice parameters	Cell volume	Crystal size (nm)
Eu^{3+} 2 at.%	5.344	157.76	9.45
Eu^{3+} 4 at.%	5.835	157.79	11.21
Eu^{3+} 6 at.%	5.423	157.81	13.54

where the crystal size is denoted by D, λ is the X-ray wavelength, maximum diffraction angle is θ, and K is a constant (0.9). No similar diffraction peaks were observed for samples that contained Eu, indicating that the Eu ions are properly integrated into the ZnS lattice. The full-width half maximum (FWHM) of the diffraction peaks broadens as the doping concentration increases due to the quantum size effect. Dopants frequently serve as nucleation sites during the synthesis of nanoparticles, lowering the energy needed for the formation of the first particle and promoting the growth of a higher number of smaller ones [20]. The dopant atoms' drag force against the increase in crystallite size is the main cause of it. It ultimately results in a smaller crystallite size. The particle size was found to be 9–13 nm. Calculation of crystal size, cell volume, and lattice parameters from XRD are shown in Table 2.

3.2 *Morphological Analysis of SEM*

Many optoelectronic applications heavily depend on microstructure and surface shape. Figure 2a shows the morphology of the as-prepared Eu-doped ZnS nanoparticles as determined by SEM. Small, uniformly sized grains with a uniform shape can be found in the SEM image of Eu-doped ZnS. Due to the Eu-doped ZnS's reduced size and increased surface energy, the sample seems to be more aggregated. From Fig. 2b, the EDX spectra, we can confirm the presence of Eu, Zn, and S. The average particle size calculated from Eu^{3+} doped ZnS 4 at.% was found to be 22 nm Fig. 2c.

3.3 *TEM*

Figure 3a shows a TEM image of Eu^{3+} doped ZnS 4 at.%. The particle size was found to be 13 nm. It agrees exactly with the results of the XRD's Scherer equation based calculation of crystallite size. The catalyst's photocatalytic activity is greatly influenced by the size of its particle. A thin layer of ZnS was laid over the surface of the Eu^{3+} doped ZnS nanocrystals. Figure 3a depicts the formation of a thin ZnS shell layer on top of the nanocrystals [21]. Reduced particle size produces a wide surface area that is ideal for high photocatalytic activity. The SAED pattern enables the exact indexing of various rings (311, 220, and 111) in the same positions.

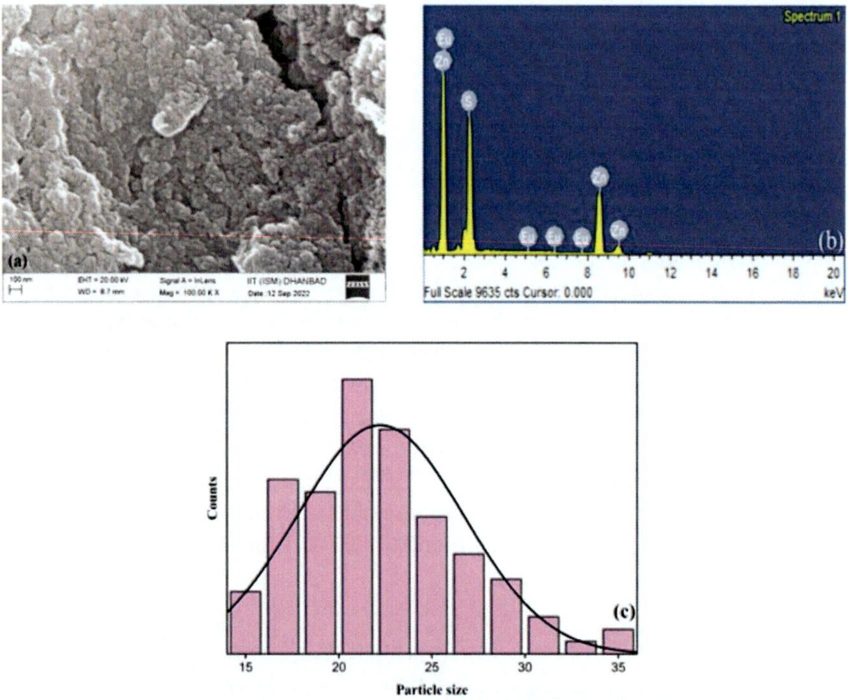

Fig. 2 **a** SEM image of Eu^{3+} doped ZnS 4 at.% **b** EDX spectra of Eu^{3+} doped ZnS **c** Average size particle size of Eu^{3+} doped ZnS 4 at.%

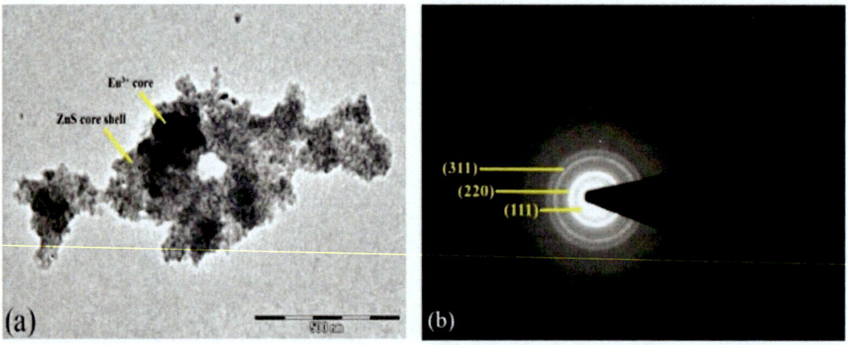

Fig. 3 **a** TEM image of Eu^{3+} doped ZnS 4 at.% **b** SAED spectra of Eu^{3+} doped ZnS

3.4 FTIR

Figure 4 shows a large absorption band at 3460 cm^{-1}, which is associated with the stretching vibration of N–H. The distinctive stretching vibration of C–H is what

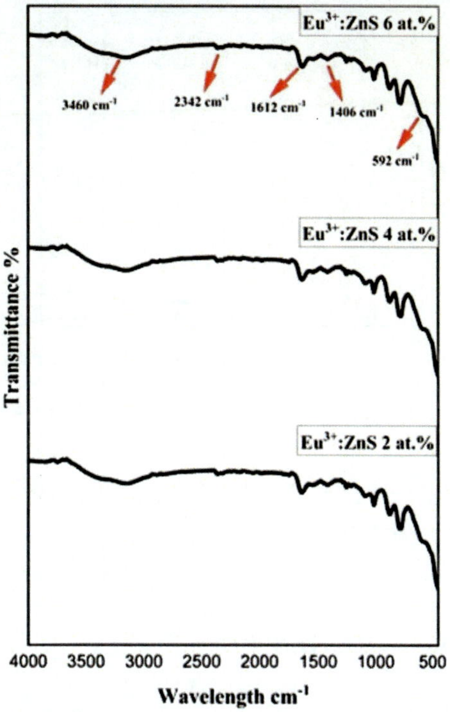

Fig. 4 FTIR spectra of Eu^{3+} 2 at%, 4 at.% and 6 at.%

causes the peak to appear at 2342 cm^{-1}. The N–H and C–H bending peaks are visible at 1612 cm^{-1} and 1406 cm^{-1}, respectively. In addition, the peak at 592 cm^{-1} belongs to the stretching vibration of Zn–S, which is also similarly reported in the literature [22].

3.5 *Photoluminescence Studies*

At room temperature, the excitation and emission properties of solid-state photoluminescence were studied. When excited at 350 nm, the major emission peaks observed are given in Fig. 5a. Different transitions of 528 nm ($^5D_0 \rightarrow {}^7F_1$), 543 nm ($^5D_0 \rightarrow {}^7F_2$), 565 nm ($^5D_0 \rightarrow {}^7F_3$), 572 nm ($^5D_0 \rightarrow {}^7F_4$), 603 nm ($^5D_0 \rightarrow {}^7F_5$), and 636 nm ($^5D_0 \rightarrow {}^7F_6$) were obtained. The strongest was the induced electric dipole transition $^5D_0 \rightarrow {}^7F_2$, which is especially susceptible to the chemical bond in close proximity to an ion of europium (III) and emits red when exposed to UV light [23, 24]. Whereas the electric dipole transition ($^5D_0 \rightarrow {}^7F_2$) is relatively strong, the magnetic dipole transition ($^5D_0 \rightarrow {}^7F_1$) is less susceptible to the coordinated environment. When emitted at 510 nm, the excitation peaks of europium are observed at 362 nm

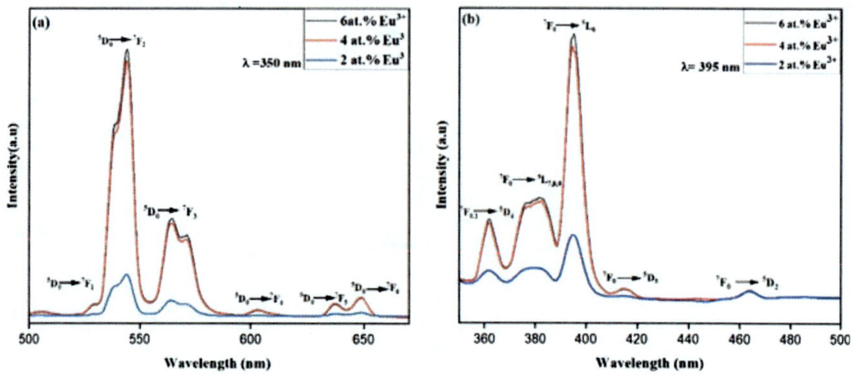

Fig. 5 **a** Emission spectra of Eu^{3+} doped ZnS **b** Excitation spectra of Eu^{3+} doped ZnS

($^7F_0 \rightarrow {}^5D_4$), 377, 380, 382 nm ($^7F_0 \rightarrow {}^5L_{7,8,9}$), 395 nm ($^7F_0 \rightarrow {}^5L_6$), 415 nm ($^7F_0 \rightarrow {}^5D_3$), 463 nm ($^7F_0 \rightarrow {}^5D_2$).

3.6 CIE Analysis

The chromaticity diagram from the Commission International de L'Eclairage (CIE) shows that the coordinates are highly helpful in defining the perceived colour of a phosphor's emission. The emission spectra-derived colour coordinates and their accompanying co-related colour temperatures are provided in Table 3. McCamy's equation was used to determine the co-related colour temperatures (CCT) [25] for Eu^{3+} doped ZnS.

$$CCT = -437n3 + 3601n2 - 6861n + 5514.31 \, and \, n = (x - xe)/(y - ye)$$

The determined colour coordinates are represented by x and y in this case. Figure 6 depicts the CIE chromaticity diagram for the obtained Eu^{3+} doped ZnS doped with various concentrations.

Table 3 CCT colour coordinates

Sample	Calculated colour coordinates		CCT
	x	y	
2 at.% Eu^{3+}	0.351	0.372	4854
4 at.% Eu^{3+}	0.349	0.388	4965
6 at.% Eu^{3+}	0.335	0.390	5417

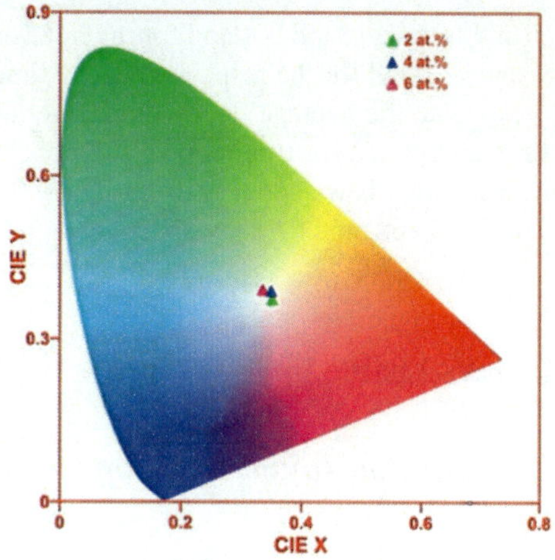

Fig. 6 CIE chromaticity diagrams of the Eu^{3+} doped ZnS (2, 4, 6 at.%)

3.7 *Lifetime Study*

Eu^{3+} doped ZnS nanoparticles have been assessed based on their luminescence decay curves. The excitation and emission wavelengths were fixed at 350 and 510 nm for each concentration, respectively, to measure the decay curves. Figure 7 displays the calculated decay curves. With the use of the bi-exponential equation, all observed decay curves are successfully fit [26].

$$I = I_1 \exp(-t/\tau_1) +_{I2} \exp(-t/\tau_2)$$

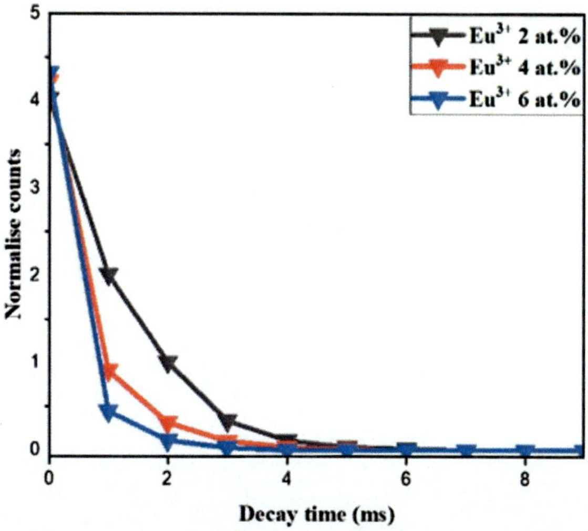

Fig. 7 Bi-exponential fitting of the photoluminescence decay curve of Eu^{3+} doped ZnS

In this case, I_1 and I_2 stand for the intensities at two distinct time points, whereas τ_1 and τ_2 stand for the respective decay times. The following equation can be used to calculate the average decay time. Average lifetimes of 1.41, 1.69, and 1.79 ms were determined for Eu^{3+} doped ZnS concentrations of 2 at.%, 4 at.%, and 6 at.%, respectively. Their goodness of fitting (R^2) for the aforementioned concentrations is 0.958, 0.968, and 0.983, respectively.

4 UV–Visible Spectroscopy

4.1 Effect of Initial Dye Concentration

At various concentrations between 5 and 30 ppm, the impact of the initial dye concentration on the photodegradation of methyl red was studied. The degradation at varied initial dye concentrations is shown in Fig. 8. When the initial dye concentration was raised, it was noticed that the percent degradation gradually dropped. This could be a result of the reaction mixture's colour getting darker and darker as the dye's initial concentration rises. This limits the ability of light to reach the catalyst's surface. Due to the constant light intensity and irradiation time, the relative amount of (OH• and $O{\bullet}_2^-$) production on the catalyst surface does not increase. Since the light photons are largely absorbed by the dye molecules and are not able to reach the catalyst surface, their concentrations will decrease as the dye concentration increases [27]. As a result, the dye's ability to degrade becomes less effective as dye concentration rises. We found that 15 ppm was the optimum dye concentration.

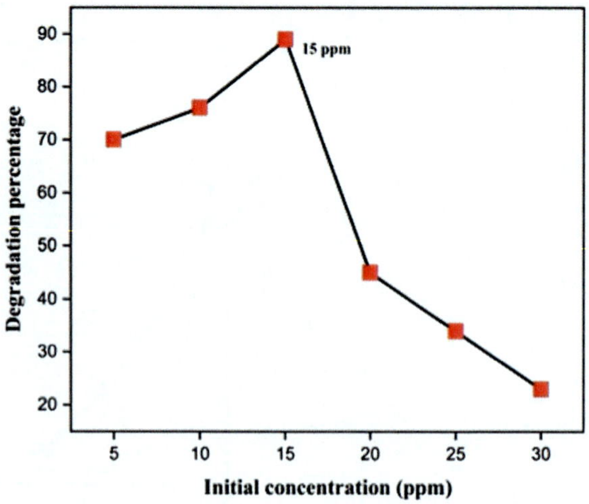

Fig. 8 Effect of initial concentration of dye (in ppm)

Fig. 9 Effect of amount of catalyst (mg/L)

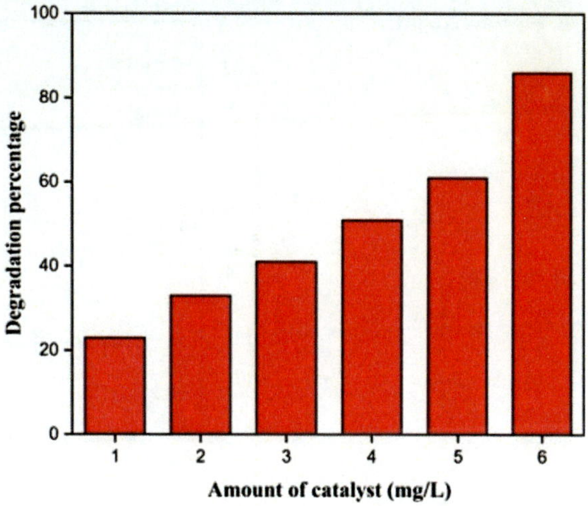

4.2 Effect of Amount of Catalyst

By irradiating dye solution (15 ppm), various catalytic doses ranging from 1 to 6 mg/L, the impact of photocatalyst concentration on dye degradation was examined. Figure 9 displays a plot of dye degradation percentage versus catalytic dosage. With an increase in catalytic dosage, the percentage of degradation rises. The dye was shown to decolorize most when exposed to radiation for 180 min at a catalytic dose of 6 mg/L. Hence, a 6 mg/L optimal catalytic dose was chosen (Fig. 9).

4.3 Effect of pH

Figure 10 displays the rate of methyl red degradation as a function of time and pH. This anionic dye degraded more quickly under photochemical conditions at pH 3 as shown in the figure. The electrostatic interactions between a semiconductor surface, substrate, and charged radicals during a photocatalytic oxidation are highly dependent on the pH of the solution. The surface of the catalyst has a higher positive charge in an acidic environment, which facilitates dye molecule adsorption. Acidic media also reduces the amount of e/h^+ recombination [28]. The catalyst surface becomes negatively charged when the pH rises, and dye molecules may repel one another, resulting in low molecule adsorption and a reduction in the efficiency of degradation. Hence, raising pH maximizes photocatalytic performance.

Fig. 10 Effect of pH

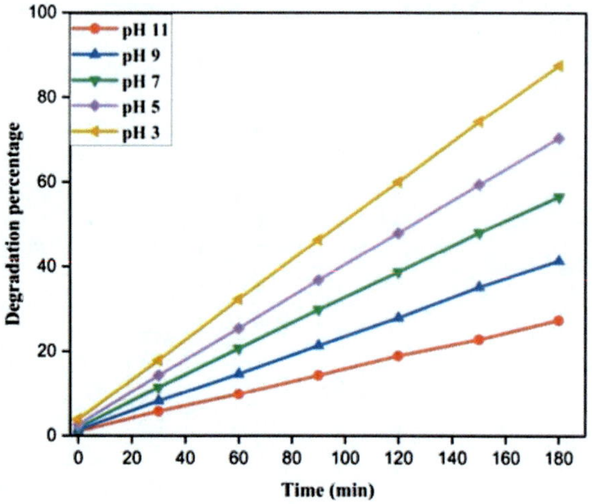

Methyl red was degraded following the optical parameters such as the effect of catalytic dosage, intial concentration, and amount of catalyst. Using 15 ppm of initial concentration, 6 mg/L of catalytic dosage, and pH 3, methyl red was degraded within 180 min. The percentage degradation of dye was calculated from the following equation:

$$Percentage\, of\ Degradation = A_0 - A_t/A_0 \times 100$$

A_0 is the initial dye absorption before degradation and the absorbance of dye is A_t. The degradation percentage of methyl red was found to be 90% Fig.s 11.

Fig. 11 Degradation of methyl red within 180 min

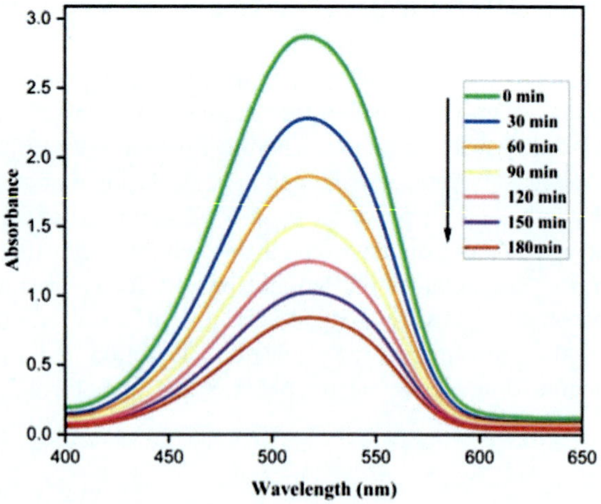

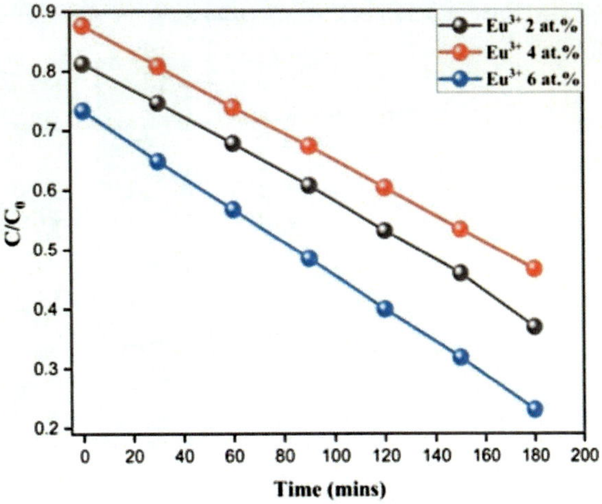

Fig. 12 Kinetic study of Eu^{3+} doped ZnS 2, 4, 6 at.%

Table 4 Kinetic study of Eu^{3+} doped ZnS

Sample	Rate constant $k(\mathrm{min}^{-1})$	R^2	DE%
Eu^{3+} 2 at.%	4.04×10^{-3}	0.958	62
Eu^{3+} 4 at.%	3.12×10^{-2}	0.968	87
Eu^{3+} 6 at.%	3.89×10^{-2}	0.983	90

4.4 *Kinetic Study*

Kinetic studies can be used to clarify the process and efficacy of the photodegradation. A linear fitting of ln C_0/C_t versus t was used to apply the first-order kinetic model [29]:

$$In\, C_0/C_t = kt$$

where (k) is the response rate constant and (t) is the time. As a result, it is evident that under UV light illumination, catalysts that are 6 at.% doped have better photodegradation efficiencies than catalysts that are 2 at.% or 4 at.%. Kinetic study of Eu^{3+} doped ZnS 2, 4, 6 at.% is shown in Fig. 12. Kinetic study of Eu^{3+} doped ZnS is tabulated in Table 4.

4.5 *Stability Analysis*

After the first cycle was finished, the catalyst was carefully removed by centrifugation, washed with ethanol/water, and dried at 120 °C in order to conduct the stability

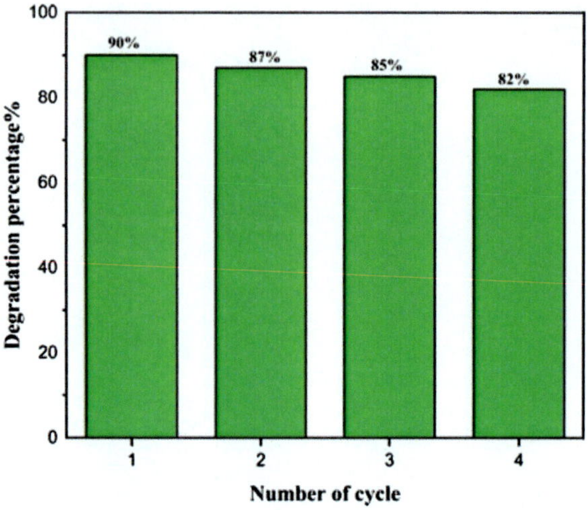

Fig. 13 Number of cycles performed

study. By utilizing a 180-min dye solution degradation process, the catalyst regenerated was used to assess its stability and reusability. The same amount of catalyst and dye was used for each cycle. The degradation of methyl red was 82% after the fourth cycle. The number of cycles performed is shown in Fig. 13.

5 Conclusion

We can conclude that Eu^{3+} doped ZnS nanoparticles of 2at.%, 4 at.%, and 6 at.% were synthesized by using a reflux method. The XRD studies show the particles are nanosized. Different transitions of 528 nm ($^5D_0 \rightarrow {}^7F_1$), 543 nm ($^5D_0 \rightarrow {}^7F_2$), 565 nm ($^5D_0 \rightarrow {}^7F_3$), 572 nm ($^5D_0 \rightarrow {}^7F_4$), 603 nm ($^5D_0 \rightarrow {}^7F_5$), and 636 nm ($^5D_0 \rightarrow {}^7F_6$) were obtained from the photoluminescence studies. According to the CIE diagram, all of the prepared samples gave out a nearly white colour that may be employed in optical devices. Methyl red can be degraded successfully by employing 6 mg/L of the catalyst, initial concentration of 15 ppm, and at pH 3. The degradation percentage was found to be 90%, and it also follows the first-order kinetics. Therefore, the synthesized Eu^{3+} doped ZnS nanoparticles can be used for removing toxic chemicals present in the environment.

Acknowledgements Not applicable.

Conflict of Interest No potential conflict of interest was declared by the author.

References

1. Elango G, Roopan SM (2016) EffiCacy of SnO 2 Nanoparticles toward Photocatalytic Degradation of Methylene Blue Dye. J Photochem Photobiol B Biol 155:34–38
2. Shabani M, Haghighi M, Kahforoushan D, Haghighi A (2018) Mesoporous-mixed-phase of hierarchical bismuth oxychlorides nanophotocatalyst with enhanced photocatalytic application ın treatment of antibiotic effluents. J Clean Prod
3. Amini M, Ashrafi M (2016) Photocatalytic degradation of some organic dyes under solar light ırradiation using TiO 2 and ZnO nanoparticles. 1(1):79–86
4. Hasanpour M, Hatami M (2020) Photocatalytic performance of aerogels for organic dyes removal from wastewaters: review study. J Mol Liq 309:113094
5. Pavithra KG, Kumar PS, Jaikumar V, Rajan PS (2019) Removal of colorants from wastewater : a review on sources and treatment. J Ind Eng Chem
6. Lai X, Guo R, Xiao H, Lan J, Jiang S, Cui C, Ren E (2019) Rapid microwave-assisted biosynthesized silver/dandelion catalyst with superior catalytic performance for dyes degradation. J Hazard Mater 371:506–512
7. Hussain W, Badshah A, Hussain RA (2017) Graphical abstract SC. Mater Chem Phys
8. Hussain W, Malik H, Bahadur A, Hussain RA, Shoaib M, Iqbal S, Hussain H (2018) Synthesis and characterization of cds photocatalyst with different morphologies: visible light activated dyes degradation study. Kinet Catal 59(6):710–719
9. Anwer H, Mahmood A, Lee J, Kim K, Park J, Yip ACK (2019) Photocatalysts for degradation of dyes in ındustrial effluents : opportunities and challenges. Nano Res 12(1)
10. Das TK, Remanan S, Ghosh S, Ghosh SK, Das NC (2021) Efficient synthesis of catalytic active silver nanoparticles ılluminated cerium oxide nanotube : A mussel ınspired approach. Environ Nanotechnol Monit Manag 15:100411
11. Hitam CNC, Jalil AA (2019) A review on exploration of Fe_2O_3 photocatalyst towards degradation of dyes and organic contaminants. J Environ Manage 258:110050
12. Preet K, Gupta G, Kotal M, Kansal SK, Salunke DB (2019) Mechanochemical synthesis of a new triptycene-based imine-linked covalent organic polymer for degradation of organic dye. Cryst Growth Des 19:2525–2530
13. Tanveer M, Cao C, Aslam I, Ali Z, Idrees F, Tahir M, Khan WS, Butt FK, Mahmood A (2014) Effect of the morphology of CuS upon the photocatalytic degradation of organic dyes. RSC Adv 4(108):63447–63456
14. Yan X, Wu Z, Huang C, Liu K, Shi W (2016) Author' s accepted manuscript. Ceram Int
15. Jrad A, Naouai M, Abdallah A, Ammar S, Turki-Kamoun N (2021) Doping ZnS films with cobalt: structural, compositional, morphological, optical, electrical, magnetic and photocatalytic properties. Phys B Condens Matter 603:412776
16. Annalakshmi M, Kumaravel S, Chen SM, Balasubramanian P, Balamurugan TST (2020) A straightforward ultrasonic-assisted synthesis of zinc sulfide for supersensitive detection of carcinogenic nitrite ıons in water samples. Sens Actuat B Chem 305:127387
17. Sharma K, Raizada P, Hasija V, Singh P, Bajpai A, Nguyen VH, Rangabhashiyam S, Kumar P, Nadda AK, Kim SY, Varma RS, Le TTN, Van Le Q (2021) ZnS-based quantum dots as photocatalysts for water purification. J Water Process Eng 43:102217
18. Lin Y-F, Chang Y-H, Chang Y-S, Tsai B-S, Li Y-C (2006) Luminescent properties of trivalent praseodymium-doped barium zinc sulfide. J Electrochem Soc 153(6):G543
19. Kaur J, Sharma M, Pandey OP (2015) Structural and optical studies of undoped and copper doped zinc sulphide nanoparticles for photocatalytic application. Superlattices Microstruct 77:35–53
20. Murugadoss G, Prakash J, Kumar MR, Alothman AA, Habila MA, Peera SG (2022) Controlled synthesis of europium-doped SnS quantum dots for ultra-fast degradation of selective ındustrial dyes. Catalysts
21. Zhang Y, Das GK, Xu R, Thatt T, Tan Y (2009) Tb-doped ıron oxide : bifunctional fluorescent and magnetic nanocrystals. 3696–3703

22. Online VA, Gong X, Liu Z, Yan D, Zhao H, Li N, Zhang X, Du Y (2015) Nanocrystals constructed by co-thermal solution phase. 3(Cd):3902–3907
23. Monteiro JHSK, De Bettencourt-Dias A, Mazali IO, Sigoli FA (2015) The effect of 4-halogenobenzoate ligands on luminescent and structural properties of lanthanide complexes: experimental and theoretical approaches. New J Chem 39(3):1883–1891
24. Hussain S, Chen X, Harrison WTA, Elsegood MRJ, Ahmad S, Li S, Muhammad S, Awoyelu D (2019) Synthesis, crystal structures and photoluminescent properties of one-dimensional Europium(III)- and Terbium(III)-Glutarate coordination polymers, and their applications for the sensing of Fe3+ and nitroaromatics. Front Chem 7:1–14
25. Naorem RS, Singh NP, Singh NM (2022) Photoluminescence properties of Tb3+ doped BiPO4 nanophosphor synthesised by co-precipitation method using ethylene glycol. Mater Technol 37(8):610–617
26. Premjit Singh N, Rangeela Devi Y, Rajmuhon Singh N, Mohondas Singh N (2019) Synthesis of Tb3+ ion doped ZnWO4 phosphors and investigation of their photoluminescence properties: concentration effect. Bull Mater Sci 42(3):17–20
27. Sahoo C (2005) Photocatalytic degradation of methyl red dye in aqueous solutions under UV ırradiation using Ag+doped TiO_2. 181:91–100
28. Thota S, Tirukkovalluri SR, Bojja S (2014) Visible light ınduced photocatalytic degradation of methyl red with codoped Titania. 2014
29. Naorem RS, Singh NP, Singh NM (2022) Synthesis of SnS2 photocatalysts and investigation of their photocatalytic activities under sun light irradiation. Adv Mater Process Technol 8(4):4720–4730

Chapter 3
Buckling of Stiffened Carbon Nanotube-Reinforced Composite (CNTRC) Plates

S. M. Shiyekar and Ranjit Wadkar

Abstract This article explores the buckling phenomenon in UG-CNTRC plates exposed to in-plane loading. The analytical modelling approach employed a Reissner–Mindlin (FOST), and a numerical computational coding, ABAQUS, was utilized to examine the performance of the (UD-CNTRC) plate-subjected buckling. Effective material properties of SWCNT were obtained by applying RM. The validity of the results from the computational tool was validated by comparison with existing literature and analytical FOST. An analysis was conducted on different variables, including the side-to-thickness ratio, CNT volume fraction, loading as well as boundary specification, to investigate their impacts on the stability performance of plates. The research further examined stiffened CNTRC plates, and the findings were presented.

Keywords Analytical solution · Uni-directional · Carbon nanotube reinforced composite · Plate · First-order shear deformation theory · ABAQUS

1 Introduction

Composites are manufactured by blending two or more disparate materials to create a unique material with enhanced properties when compared to the separate components. Nanocomposites refer to materials in which nanoparticles are added to enhance a specific property. The incorporation of nanoparticles can enhance material attributes such as endurance and conductivity. The use of nanoparticles in structural material components results in a high strength-to-weight ratio.

S. M. Shiyekar (✉)
Department of Civil Engineering, D Y Patil College of Engineering, Akurdi, Pune, India
e-mail: smshiyekar@dypcoeakurdi.ac.in

R. Wadkar
Department of Civil Engineering, Rajarambapu Institute of Technology, Rajaramnagar, Maharashtra, India

V. Bindhu et al. (eds.), *Proceedings of Sixth International Conference on Inventive Material Science Applications*, Advances in Sustainability Science and Technology,
https://doi.org/10.1007/978-981-99-4189-6_3

Innovation by Iijima [1] in 1991, attracted researchers and they referred to it as the "material for the twenty-first century" [2]. CNTs have nanodimensions comprised of rolled layers of graphite and can be arrived in different shapes such as "zigzag," "chiral," and "armchair". SWCNT ("Single-wall carbon nanotube reinforced composite plates") or MWCNT (multi-wall) tubes may be formed considering the layer of graphite is rolled along. Previous research [3–5] has demonstrated that CNTs exhibit excellent properties, such as exceptional stiffness and strength. Researchers [6, 7] have conducted experiments and simulations [8] to demonstrate the load-taking abilities of CNTs in composites. Despite the great potential of CNT-based composites, there are also significant challenges in their development. Researchers have used various methods, such as RVE, FEM, and MD simulations, to evaluate actual material attributes related to CNT-based composites and also to examine their thermo-mechanical properties. Studies have also explored the stress–strain performance of CNT composites [9–13].

The studies mentioned above provide useful insights into the performance of CNTRCs in different structures. Wuite and Adali [14] examined CNTRC beams. Vodenitcharova and Zhang [15] explored the flexure local buckling and pure bending of nanobeams strengthened by an SWNT. Lei et al. [16] presented a report on the stability of functionally graded and uniformly distributed (FG and UD) CNTRC plates. Similarly, Malekzadeh and Shojaee [17] researched the buckling of CNTRC thick plates. A stability investigation of FG nanocomposite bi-directional plates reinforced by SWCNTs was carried out by Jafari et al. [18]. To ascertain the SWCNT's material characteristics, they used the "extended rule of mixture" (ERM). Formica et al. [19] used the "classical laminated plate theory" (CLPT) and "third-order theory" (TOT) to study the dynamics of CNTRC plates. The CLPT was used by Arani et al. [20] to illustrate the buckling of CNTRC bi-directional plates. This plate is subject to a variety of boundary constraints. The MT (Mori–Tanaka) approach is used to assess the efficient elastic modulus of composites containing straight nanotubes using the "finite element method" (FEM) tool. Shen [21–23] examined the post-buckling properties of cylindrical FG CNTRC shells exposed to axial compression at temperature. Christian Mittelstedt [24] presented the buckling of orthotropic and isotropic plates with stiffeners. Kant and Swaminathan [25] investigated the stability of laminated composite and sandwiches. Timoshenko and Gere [26] conducted a buckling analysis of an isotropic plate. Kurkute and Shiyekar [27] presented the buckling of laminates using higher-order theory. Mayandi and Jeyaraj [28] discussed flexure, buckling and free vibrations of FG CNT beams under thermal loads. Shiyekar and Lavate [29] analyzed FG plates using ABAQUS UMAT.

The present study focuses on UD-CNTRC plate buckling under in-plane forces with various boundary parameters. The FEM and the Mindlin [30] (FOST) are used. The commercial program ABAQUS [31] is employed as the computational tool. The study includes three categories of plates: isotropic plate, UD-CNTRC plates, and CNTRC plates with stiffeners along the longitudinal direction. The influences of several factors, including plate aspect ratio, CNT volume fraction, as well as boundary limits on the CNTRC plate buckling strength, are studied.

2 Modeling of CNTRC Material

2.1 *Extended Rule of Mixture (ERM)*

The CNTRC plates are produced through a fusion of SWCNTs and polymer matrix, where the attributes of SWCNTs vary depending on their handedness size and thermal conditions. The present study considers UD-CNTRC: "Carbon nanotube-reinforced composite" plates with '*h*' thickness, '*a*' length, and '*b*' width (Fig. 1), and the MT methodology or the RM (Rule of Mixture) will be used to identify the real material characteristics of these composites. The RM involves introducing CNT efficiency parameters, and assuming that CNTs are distributed uniformly across the UD-CNTRC plate thickness, as shown in (Fig. 1). Volume fraction of CNT is given as

$$V_{CNT} = V^{*}_{CNT}(\mathrm{UD}\ \cdot\ \mathrm{CNTRC}) \tag{1}$$

where

$$V^{*}_{CNT} = \frac{w_{CNT}}{w_{CNT} + \left(\frac{\rho_{CNT}}{\rho_m}\right) - \left(\frac{\rho_{CNT}}{\rho_m}\right) w_{CNT}} \tag{2}$$

where the CNT mass fraction in the composite plate is given by w_{CNT}. The CNT and matrix density is denoted by ρ_{CNT} and ρ_m. The ρ density and υ_{12} Poisson's ratio of the nanocomposite plates are computed as

$$\upsilon_{12} = V_{CNT}\upsilon_{12}^{CNT} + V_m\upsilon_m \tag{3}$$

$$\rho = V_{CNT}\rho_{CNT} + V_m\rho_m \tag{4}$$

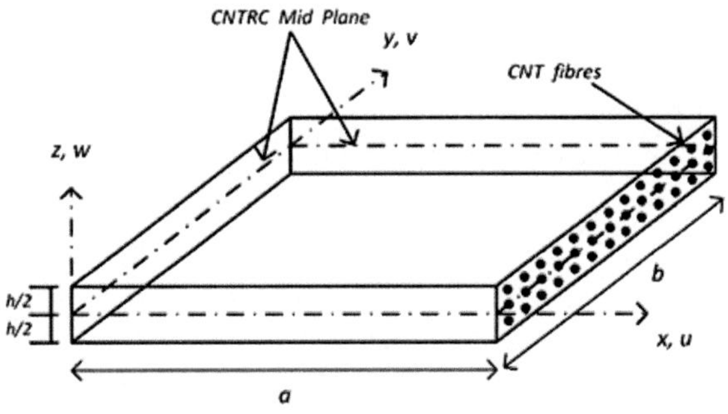

Fig. 1 Configuration of the uniformly distributed carbon nanotube reinforced composite plates (UD CNTRC)

where Poisson's ratio υ_m and υ_{CNT} of matrix and CNT correspondingly. ERM can be utilized to find the actual material characteristics of plates. By incorporating the CNT efficiency factor, the real material characteristics of the composites will be computed using this approach.

$$E_{11} = \eta_1 V_{CNT} E_{11}^{CNT} + V_m E^m \tag{5}$$

$$\frac{\eta_2}{E_{22}} = \frac{V_{CNT}}{E_{22}^{CNT}} + \frac{V_m}{E^m} \tag{6}$$

$$\frac{\eta_3}{G_{12}} = \frac{V_{CNT}}{G_{12}^{CNT}} + \frac{V_m}{G^m} \tag{7}$$

where E_{11}^{CNT}, E_{22}^{CNT} are the Modulus of Elasticity of along the length and normal directions, respectively, and G_{12}^{CNT} are shear modulus of CNTs. CNT efficiency factor (h_i) was presented to find out scale-dependent material attributes. It may be used to match the effective CNTRC characteristics discovered by the Mori–Tanaka simulation. The matrix, as well as CNT, volume is denoted by V_m and V_{CNT} correspondingly. The addition of these fractions of phases equals unity

$$V_{CNT} + V_m = 1 \tag{8}$$

3 Theoretical Formulations

FOST is used to analyse buckling in the CNTRC plate. A complete analytical solution and formulation are revealed for a simply supported CNTRC plate. The plate geometry is side 'a' is on the x-axis, 'b' is on the y-axis, and the plate thickness is represented by 'h' and matches with the 'z' direction. Figure 1 depicts that the mid-plane is situated at a distance of +h/2 from the top and −h/2 from CNTRC plate bottom. The mid-plane positive displacement set with the (x-y-z) axes is utilized with the reference cartesian axis system. The CNTRC plate stability is examined using the given below displacement domain.

$$\begin{aligned} u(x, y, z) &= u_0(x, y) + z\theta_x(x, y) \\ v(x, y, z) &= v_0(x, y) + z\theta_y(x, y) \\ w(x, y) &= w_0(x, y) \end{aligned} \tag{9}$$

This strain–displacement connections for small displacements generally may be expressed as follows according to the linear theory of solid mechanics.

$$\begin{Bmatrix} \varepsilon_x \\ \varepsilon_y \\ \varepsilon_z \\ \gamma_{xy} \\ \gamma_{xz} \\ \gamma_{yz} \end{Bmatrix} = \left\{ \frac{\partial u}{\partial x}\ \frac{\partial v}{\partial y}\ \frac{\partial w}{\partial z}\ \frac{\partial u}{\partial y} + \frac{\partial v}{\partial x}\ \frac{\partial u}{\partial z} + \frac{\partial w}{\partial x}\ \frac{\partial v}{\partial z} + \frac{\partial w}{\partial y} \right\}^T \tag{10}$$

Strains are obtained according to the displacement field

$$\begin{aligned} &\varepsilon_x = \frac{\partial u_0}{\partial x} + z\frac{\partial \theta_x}{\partial x}, \varepsilon_y = \frac{\partial v_0}{\partial y} + z\frac{\partial \theta_y}{\partial y}, \varepsilon_z = 0 \\ &\gamma_{xy} = \frac{\partial u_0}{\partial y} + \frac{\partial v_0}{\partial x} + z\left(\frac{\partial \theta_x}{\partial y} + \frac{\partial \theta_y}{\partial x}\right), \gamma_{xz} = \theta_x + z\frac{\partial w_0}{\partial x} \\ &\gamma_{yz} = \theta_y + z\frac{\partial w_0}{\partial y} \end{aligned} \tag{11}$$

The material constitutional association is written for composite material as.

$$\begin{Bmatrix} \sigma_x \\ \sigma_y \\ \tau_{xy} \\ \tau_{xz} \\ \tau_{yz} \end{Bmatrix} = \begin{bmatrix} C_{11} & C_{12} & 0 & 0 & 0 \\ C_{12} & C_{22} & 0 & 0 & 0 \\ 0 & 0 & C_{44} & 0 & 0 \\ 0 & 0 & 0 & C_{55} & 0 \\ 0 & 0 & 0 & 0 & C_{66} \end{bmatrix} \begin{Bmatrix} \varepsilon_x \\ \varepsilon_y \\ \gamma_{xy} \\ \gamma_{xz} \\ \gamma_{yz} \end{Bmatrix} \tag{12}$$

where $C_{11} = \dfrac{E_{11}}{1 - \upsilon_{12}\upsilon_{21}}, C_{12} = \dfrac{\upsilon_{21}E_{11}}{1 - \upsilon_{12}\upsilon_{21}}, C_{22} = \dfrac{E_{22}}{1 - \upsilon_{12}\upsilon_{21}},$.
$C_{44} = G_{12}, C_{55} = G_{13}, C_{66} = G_{23}$

For stress analysis, the governing differential equilibrium forms have resulted from the "principle of minimum potential energy". The mathematical equation is expressed as follows:

$$\int (\delta U - \delta W)dV + \int \left(\sigma'\delta\varepsilon'_x + \sigma'\delta\varepsilon'_y + \tau'_{xy}\delta\gamma'_{xy}\right)dV = 0 \tag{13}$$

where the strain energy due to deformations is denoted by U, the work performed by outside loads W, the second term in the above equation is potential energy because of given in-plane compressive and shear load (i.e., σ'_x, σ'_y, τ'_{xy}). The midplane strains caused by transverse displacement are ε'_x, ε'_y, γ'_{xy} and the variational symbol is given by δ. The final expression will be after the necessary quantities of energy have been entered into the equation.

$$\left[\int_{-\frac{h}{2}}^{\frac{h}{2}}\int_A \left(\sigma_x\delta\varepsilon_x + \sigma_y\delta\varepsilon_y + \tau_{xy}\delta\gamma_{xy} + \tau_{yz}\delta\gamma_{yz} + \tau_{xz}\delta\gamma_{xz}\right)dAdz - \int_A p_z^+\delta w^+ dA\right]$$
$$+\int_{-\frac{h}{2}}^{\frac{h}{2}}\int_A \left(\sigma'\delta\varepsilon'_x + \sigma'\delta\varepsilon'_y + \tau'_{xy}\delta\gamma'_{xy}\right)dAdz = 0 \tag{14}$$

where the through-thickness displacement of any point is denoted by $w^+ = w_0$ and the transverse load given to the CNTRC plate top is represented by p_z^+. By applying the above Eqs. (9) to (13) in (14) and integrating the equations by parts, the following relationships of equilibrium are obtained.

$$\begin{aligned}
&\delta u_0 : \frac{\partial N_x}{\partial x} + \frac{\partial N_{xy}}{\partial y} = 0, \delta v_0 : \frac{\partial N_y}{\partial y} + \frac{\partial N_{xy}}{\partial x} = 0,\\
&\delta w_0 : \frac{\partial Q_x}{\partial x} + \frac{\partial Q_y}{\partial y} + \left(\frac{\partial w_0}{\partial x}\frac{\partial \overline{N}_y}{\partial x}\right) + \left(\frac{\partial w_0}{\partial y}\frac{\partial \overline{N}_y}{\partial y}\right)\\
&+\overline{N}_x\frac{\partial^2 w_0}{\partial x^2} + \overline{N}_y\frac{\partial^2 w_0}{\partial y^2} - \left(\frac{\partial^2 \overline{N}_{xy}}{\partial x \partial y}\right) = 0,\\
&\delta\theta_x : \frac{\partial M_x}{\partial x} + \frac{\partial M_{xy}}{\partial y} - Q_x = 0, \delta\theta_y : \frac{\partial M_y}{\partial y} + \frac{\partial M_{xy}}{\partial x} - Q_y = 0
\end{aligned} \tag{15}$$

where the stress resultants are presented by

$$\begin{bmatrix} M_x \\ M_y \\ M_{xy} \end{bmatrix} = \int_{-\frac{h}{2}}^{+\frac{h}{2}} \begin{Bmatrix} \sigma_x \\ \sigma_y \\ \tau_{xy} \end{Bmatrix} [z]dz, \begin{bmatrix} Q_x \\ Q_y \end{bmatrix} = \int_{-\frac{h}{2}}^{+\frac{h}{2}} \begin{Bmatrix} \tau_{xz} \\ \tau_{yz} \end{Bmatrix} dz, \begin{bmatrix} N_x \\ N_y \\ N_{xy} \end{bmatrix} = \int_{-\frac{h}{2}}^{+\frac{h}{2}} \begin{Bmatrix} \sigma_x \\ \sigma_y \\ \tau_{xy} \end{Bmatrix} dz,$$

and

$$\left[\overline{N}_x\ \overline{N}_y\ \overline{N}_{xy}\right] = \int_{-h/2}^{+h/2} \left[\sigma_{x'}\ \sigma_{y'}\ \tau_{xy'}\right]dz \tag{16}$$

The resultants are related to the displacement and strain relationship and may be further written in the given equation as displacements

$$
\begin{aligned}
&N_x = h\left(Q_{12}\frac{\partial v_0}{\partial y} + Q_{11}\frac{\partial u_0}{\partial x}\right), N_y = h\left(Q_{22}\frac{\partial v_0}{\partial y} + Q_{12}\frac{\partial u_0}{\partial x}\right),\\
&N_{xy} = hQ_{44}\left(\frac{\partial u_0}{\partial y} + \frac{\partial v_0}{\partial x}\right), M_x = \frac{h^3}{12}\left(Q_{12}\frac{\partial \theta_y}{\partial y} + Q_{11}\frac{\partial \theta_x}{\partial x}\right),\\
&M_y = \frac{h^3}{12}\left(Q_{22}\frac{\partial \theta_y}{\partial y} + Q_{12}\frac{\partial \theta_x}{\partial x}\right), M_{xy} = \frac{h^3}{12}Q_{44}\left(\frac{\partial \theta_x}{\partial y} + \frac{\partial \theta_y}{\partial x}\right),\\
&Q_x = hQ_{66}\left(\frac{\partial w_0}{\partial x} + \theta_x\right), Q_y = hQ_{55}\left(\frac{\partial w_0}{\partial y} + \theta_y\right).
\end{aligned}
\tag{17}
$$

Substituting the above stress resultants in the equilibrium Eq. (15), the new set of governing mathematical forms are deduced as displacements as follows:

$$
\begin{aligned}
&hQ_{44}\frac{\partial^2 u_0}{\partial y^2} + hQ_{12}\frac{\partial^2 v_0}{\partial x \partial y} + hQ_{44}\frac{\partial^2 v_0}{\partial x \partial y} + hQ_{11}\frac{\partial^2 u_0}{\partial x^2} = 0\\
&hQ_{22}\frac{\partial^2 v_0}{\partial y^2} + hQ_{12}\frac{\partial^2 u_0}{\partial x \partial y} + hQ_{44}\frac{\partial^2 u_0}{\partial x \partial y} + hQ_{44}\frac{\partial^2 v_0}{\partial x^2} = 0\\
&hQ_{55}\frac{\partial \theta_y}{\partial y} + hQ_{55}\frac{\partial^2 w_0}{\partial y^2} + hQ_{66}\frac{\partial \theta_x}{\partial x} + hQ_{66}\frac{\partial^2 w_0}{\partial x^2} + q = 0\\
&\frac{h^3}{12}Q_{44}\frac{\partial^2 \theta_x}{\partial y^2} - hQ_{66}\frac{\partial w_0}{\partial x} + \frac{h^3}{12}Q_{12}\frac{\partial^2 \theta_y}{\partial x \partial y} + \frac{h^3}{12}Q_{44}\frac{\partial^2 \theta_y}{\partial x \partial y} + \frac{h^3}{12}Q_{11}\frac{\partial^2 \theta_x}{\partial x^2} - hQ_{66}\theta_x = 0\\
&\frac{h^3}{12}Q_{22}\frac{\partial^2 \theta_y}{\partial y^2} - hQ_{55}\frac{\partial w_0}{\partial y} + \frac{h^3}{12}Q_{12}\frac{\partial^2 \theta_x}{\partial x \partial y} + \frac{h^3}{12}Q_{44}\frac{\partial^2 \theta_x}{\partial x \partial y} + \frac{h^3}{12}Q_{44}\frac{\partial^2 \theta_y}{\partial y^2} - hQ_{55}\theta_y = 0
\end{aligned}
\tag{18}
$$

The displacement solution is determined by applying Navier's solution. Basic supported boundary limits are applied to all sides.

$$
\text{At edges } x = a, \text{ and } x = 0: v_0 = 0, w_0 = 0, \theta_y = 0, \theta_z = 0, M_x = 0, N_x = 0. \tag{19}
$$

$$
\text{At edges } y = 0, \text{ and } y = b: u_0 = 0, w_0 = 0, \theta_x = 0, \theta_z = 0, M_y = 0, N_y = 0. \tag{20}
$$

The following system may be used to express the result of the displacement quantities supporting the aforementioned boundary situations:

$$
\begin{aligned}
&u_0 = \sum_{m=1}^{\infty}\sum_{n=1}^{\infty} u_{0_{mn}} \cos(\alpha x)\sin(\beta y), v_0 = \sum_{m=1}^{\infty}\sum_{n=1}^{\infty} v_{0_{mn}} \sin(\alpha x)\cos(\beta y),\\
&w_0 = \sum_{m=1}^{\infty}\sum_{n=1}^{\infty} w_{0_{mn}} \sin(\alpha x)\sin(\beta y), \theta_x = \sum_{m=1}^{\infty}\sum_{n=1}^{\infty} \theta_{x_{mn}} \cos(\alpha x)\sin(\beta y),\\
&\theta_y = \sum_{m=1}^{\infty}\sum_{n=1}^{\infty} \theta_{y_{mn}} \sin(\alpha x)\cos(\beta y)
\end{aligned}
\tag{21}
$$

where $\alpha = m\pi/a$, and $\beta = n\pi/b$. Substituting Eqs. (21) in (18), and the components of the coefficient matrix [*X*] will be generated by obtaining the coefficients for the displacements.

$$([X]_{5X5} - \lambda[G]_{5X5}) \begin{Bmatrix} u_0 \\ v_0 \\ w_0 \\ \theta_x \\ \theta_y \end{Bmatrix}_{5X1} = \{0\} \quad (22)$$

where λ = buckling load factor. For given *m* and *n* values, the geometric stiffness matrix *[G]* refers to the stiffness matrix because of in-plane forces. The components of matrices *[G]* & *[X]* are provided in Appendix 1. The aforementioned equation may be obtained, and the essential buckling load factor is the lowest eigenvalue and the essential uniaxial load is given by the relation.

$$[\overline{N}_x]_{cr} = \lambda_{cr}[\overline{N}_x] \quad (23)$$

4 Results and Discussions

Validation of the present FOST and ABQUS is studied through the following examples. Poly Vinylene, denoted as PmPV is deemed to be the polymer matrix in this problem. The elastic properties are supposed to be $\upsilon_m = 0.34$, $E_m = 2.1$ Gpa, and $\rho_m = 1.15$ g/cm^3 at 300 K temperature. In this paper, the armchair (10, 10) SWCNTs are considered as reinforcement. The SWCNT characteristics are calculated by the MD simulation process and are presented in Table 1.

ERM is used to forecast actual material attributes of CNTRCs. Carbon nanotube efficiency variables η_j included in Eqs. (5) using (7). We consider for $V_{CNT} = 0.11$, η_1 and η_2 values are 0.149 and 0.934 and for $V_{CNT} = 0.14$, values are 0.150 and 0.941 and for $V_{CNT} = 0.17$, values are 0.149 and 1.381, correspondingly. Additionally, we suppose that $\eta_3 = \eta_2$ and $G_{23} = G_{13} = G_{12}$ are considered in the subsequent numerical examples.

Table 1 Temperature-dependent material properties of (10, 10) SWCNT (L = 9.26 nm, R = 0.68 nm, h = 0.067 nm, υ_{12}^{CNT} = 0:175)

Temperature (K)	E_{11}^{CNT} (TPa)	E_{22}^{CNT} (TPa)	G_{12}^{CNT} (TPa)	α_{11}^{CNT} (10^{-6}/K)	α_{12}^{CNT} (10^{-6}/K)
300	5.6466	7.0800	1.9645	3.4584	5.1682
500	5.5308	6.9348	1.9643	4.5361	5.0189
700	5.4744	6.8641	1.9644	4.6677	4.8943

Table 2 Buckling load for all edges simply supported square isotropic plates under uniaxial compression

Mode	SSSS isotropic plate (b/h = 10)		
	Analytical [26]	Present (ABAQUS)	% Error
1	39.4761	39.1063	0.93
2	61.6814	61.6116	0.11
3	109.656	111.3795	−1.57
4	157.9040	157.3482	0.35

4.1 Example 1: Buckling of an Isotropic Plate Under Compressive Force

To validate the present method of analytical FOST and numerical ABAQUS, comparisons are shown for isotropic plates that are simply supported on both sides and compressed by uniaxial force. Dimensions and elastic of the plate are , $a = b = 0.254$ m, $E = 10.3 \times 10^3$ MPa (3×10^6 psi), $h = 0.0508$ m, and Poisson ratio $\upsilon = 0.3$. The buckling load is normalized as $\overline{N}_{cr} = (N_{cr}a^2/D_0)$, $D_0 = Eh^3/12(1 - \upsilon^2)$. The aspect ratio is taken as 10. The critical buckling load at fundamental 4 modes from the ABAQUS gives good agreement with analytical solutions given by Timoshenko and Gere [26] and presented in Table 2 with a maximum % error of 1.57.

4.2 Example 2: Buckling of CNTRC Plate Under (SSSS) Uniaxial Compressive Force

CNTRC plate under uniaxial condition for all sides simply supported is analyzed for $b/h = 20$ and $V_{CNT} = 0.11$. Results are compared with EF Ritz [17], DQM [18], and Exact [18] and present FOST and ABAQUS in Table 3. The first four modes are observed and the buckling load is normalized as $\overline{N_{cr}} = N_{cr}b^2/Eh^3$. Uniaxial and biaxial compression on CNTRC plates is done using a side-to-thickness ratio of 10 and $V_{CNT} = 0.11$ are presented in Table 4. In the table, the literature and the present ABAQUS buckling load values are compared. The following boundary conditions are used: (1) **SSSS**–Fully supported four sides, (2) **CCCC**–Fully clamped four sides (fixed), (3) **SCSC**–Two opposite supported sides and rest two are clamped sides, (4) **SFSF**–Two opposite supported sides and rest two are free sides, (5) **CCCF**–Three clamped sides and one side free, (6) **SSSF**–Three simply supported sides and one free side.

Table 3 Comparison of different methods of the critical buckling load for simply supported plate with single CNTRC layer under uniaxial compression, $V_{CNT} = 0.11$, $b/h = 20$

Methods	Buckling load
Exact [18]	31.5258
DQM [18]	31.5258 [0]
EF-Ritz [17]	30.9076 [1.96]
Present analytical (FOST)	29.4578 [6.55]
Present (ABAQUS)	32.3200 [−2.51]

[]–% error with respect to Exact [18]

Table 4 Buckling load for all edges simply supported and all edge clamped (fixed) square CNTRC plates under uniaxial and biaxial compression with $b/h = 10$, $V_{CNT} = 0.11$. (M = Modes)

M	SSSS uniaxial			CCCC uniaxial		
	Ref. [16]	Present (ABAQUS)	% Error	Ref. [16]	Present (ABAQUS)	% Error
1	14.1073	14.2519	−1.02	25.7329	24.5720	4.51
2	23.3149	22.3948	3.94	26.2788	25.8263	1.72
3	25.6506	24.4761	4.57	29.6661	28.2214	4.86
4	27.0498	26.1428	3.35	29.7356	28.0410	5.69
M	SCSC uniaxial			SFSF uniaxial		
1	17.2424	18.0661	−4.77	12.0697	12.71	−5.30
2	24.0481	23.1510	3.73	13.0216	13.44	−3.21
3	26.3993	24.3514	7.75	20.3912	20.61	−1.07
4	27.4953	26.7247	2.80	22.4198	20.96	6.51
M	CCCF uniaxial			SSSF uniaxial		
1	22.6023	21.3624	5.48	12.3145	12.85	−4.34
2	24.8135	23.7648	4.22	16.6801	16.15	3.17
3	26.4696	25.0480	5.37	22.5393	21.68	3.81
4	26.9602	25.4476	5.61	24.1999	23.26	3.88
M	SCSC biaxial			SFSF biaxial		
1	5.8831	5.5501	5.66	9.2830	9.6385	−3.82
2	6.9772	6.7253	3.61	9.8992	10.3062	−4.11
3	7.8892	7.6581	2.92	13.1796	13.2719	−0.70
4	10.6318	10.189	4.16	13.5820	13.3671	1.58
M	CCCF biaxial			SSSF biaxial		
1	8.1646	8.2351	−0.86	3.4105	3.5380	−3.73
2	9.1246	8.9658	1.74	6.0446	5.7142	5.46
3	9.8991	9.5323	3.70	7.8034	7.6432	2.05
4	11.2443	10.8740	3.29	7.9533	8.0550	−1.27

Table 4 shows that among the six diverse boundary situations, plates that have all four edges clamped (CCCC) exhibit the highest load, while two supported edges and rest two free edges (SFSF) plates exhibit the lower load.

The results show that raising the V_{CNT} causes CNTRC plates to be stiffer, which raises the plates' buckling load parameter. The data demonstrates that when the ratio of width-to-thickness of UD-CNTRC plates rises, so do the non-dimensional buckling load characteristics.

4.3 Example 3: Buckling of Stiffened CNTRC Plates Under Compressive Pressure

The stiffening impact on the buckling behaviour of rectangular CNTRC plates (SSSS) in uniaxial compression is presented in this section. Table 1 displays that the ERM is utilized to find the actual material characteristics of CNTRCs. The geometric characteristics of the CNTRC plate and stiffener are adopted from Christian Mittelstedt [24], where the plate dimensions are a = 300 mm and b = 100 mm, and the height and thickness of both the plate and stiffener are 13.5 and 1 mm, respectively. The different stiffening systems used for CNTRC plates, include: (a) One-side stiffeners running parallel to the CNTRC plate length. (b) Middle stiffeners running parallel to the length of the CNTRC plate. (c) Two-side stiffeners running parallel to the CNTRC plate length. Table 5 illustrates the buckling load of these different stiffener positions for CNTRC plates (SSSS). The results indicate that the two-side stiffener system produces the highest buckling load, which suggests that the plate becomes stiffer in the following order: side stiffeners to the CNTRC plate length, middle stiffeners to the CNTRC plate length, and two-side stiffeners to the length of CNTRC plate. The buckling loads were also can be seen to rise as the volume percent of CNTs increased. Figure 2 displays the buckling modes of plates featuring different stiffener systems.

Table 5 Buckling load for all edges simply supported rectangle CNTRC plates under uniaxial compression

M	V_{CNT}	Without stiffeners of the CNTRC plate (ABAQUS)	One side stiffeners of the CNTRC plate (ABAQUS) (a)	Middle stiffeners of the CNTRC plate (ABAQUS) (b)	Two side stiffeners of the CNTRC plate (ABAQUS) (c)
1	0.11	12.9266	32.4392	54.3448	85.9762
2		13.0781	32.8169	54.3867	85.8039
3		13.9681	35.0879	56.3190	85.9905
4		14.1024	35.8324	28.1905	94.3638

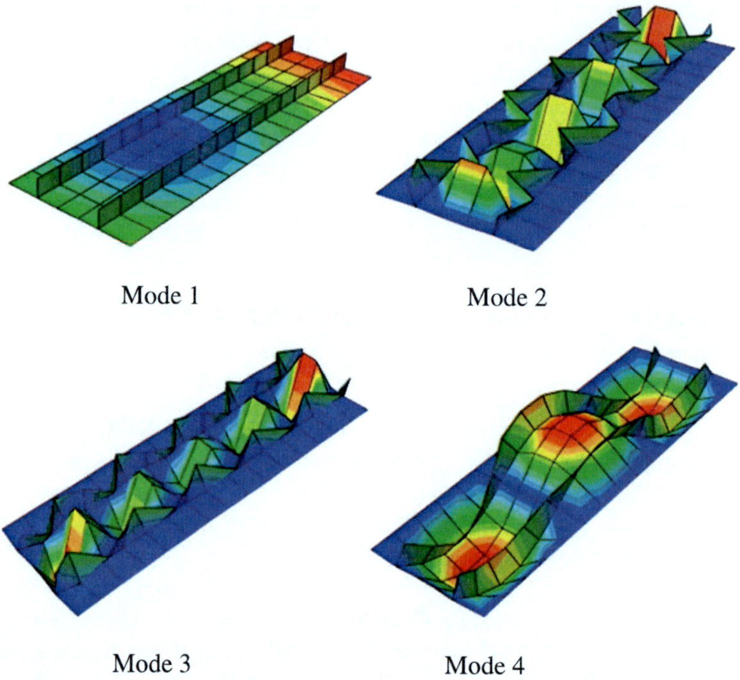

Fig. 2 Buckling modes of stiffened CNTRC plates for $V_{CNT} = 0.14$ and b/h = 10 under uniaxial compression

5 Conclusion

This article aims to estimate the buckling performance of isotropic and CNTRC plates using two methods: FOST, the analytical method and the ABAQUS computational tool. When the findings from the current approaches are compared to those from the literature, there is a high level of agreement. The ERM is applied to find the real material characteristics of the SWCNT-reinforced plates. Comprehensive case studies are performed, and CCCC, i.e., fully clamped four edges plate displays the most buckling stress, whereas SFSF, i.e., two supported edges and the rest two free edges plate exhibits the least amount of buckling load among the six alternative boundary conditions. As V_{CNT} rises, the load also increases. The buckling performance of CNTRC plates with stiffeners along the length is also investigated, and it is concluded that the plate with two stiffeners exhibits the maximum buckling load. It is important to note that various CNTRC plate configurations' buckling strength may be significantly affected by altering V_{CNT}, ratio of width-to-thickness, and aspect ratio, and the boundary and loading limits.

Appendix 1

Coefficient of Matrix [X]

$$
\begin{aligned}
&X[1,1] = Q_{11}h\alpha^2 + Q_{44}h\beta^2,\ X[1,2] = Q_{12}h\alpha\beta + Q_{44}h\alpha\beta,\ X[1,3] = 0,\\
&X[1,4] = 0,\ X[1,5] = 0,\ X[1,1] = A[1,1]\alpha^2 + B[1,1]\beta^2,\\
&X[1,2] = A[1,2]\alpha\beta + A[1,2]\alpha\beta,\ X[1,3] = 0,\ X[1,4] = 0,\ X[1,5] = 0\\
&X[2,1] = Q_{12}h\alpha\beta + Q_{44}h\alpha\beta,\ X[2,2] = Q_{44}h\alpha^2 + Q_{22}h\beta^2,\\
&X[2,3] = 0,\ X[2,4] = 0,\ X[2,5] = 0,\ X[3,1] = 0,\ X[3,2] = 0,\\
&X[3,3] = Q_{66}h\alpha^2 + Q_{55}h\beta^2,\ X[3,4] = Q_{66}h\alpha,\ X[3,5] = Q_{55}h\beta,\ X[4,1] = 0,\\
&X[4,2] = 0,\ X[4,3] = Q_{66}h\alpha,\ X[4,4] = \frac{h^3}{12}\left(Q_{11}\alpha^2 + Q_{44}\beta^2\right) + Q_{66}h,\\
&X[4,5] = \frac{h^3}{12}(Q_{12}\alpha\beta + Q_{44}\alpha\beta),\ X[5,1] = 0,\ X[5,2] = 0,\ X[5,3] = Q_{55}h\beta,\\
&X[5,4] = \frac{h^3}{12}(Q_{12}\alpha\beta + Q_{44}\alpha\beta),\ X[5,5] = \frac{h^3}{12}\left(Q_{44}\alpha^2 + Q_{22}\beta^2\right) + Q_{55}h.
\end{aligned}
$$

Coefficient of Geometric Matrix [G]

$G_{3,3} = \alpha^2$ and $G_{i,j} = 0$ for all $i, j = 1, 2.......5(i \neq j \neq 3)$.

References

1. Ijima S (1991) Helical microtubes of graphite carbon. Nature 354:56–58
2. Harris PJF (2001) Carbon nanotubes and related structures: a new material for the twenty-first century. Cambridge University Press
3. Esawi AMK, Farag MM (2007) Carbon nanotube reinforced composites: potential and current challenges. Mater Des 28:2394–2401
4. Lau AKT, Hui D (2002) The revolutionary creation of new advanced materials carbon nanotube composites. Compos Part B 33:263–277
5. Thostenson ET, Ren Z, Chou TW (2001) Advances in the science and technology of carbon nanotubes and their composites: a review. Comp Sci Technol 61:1899–1912
6. Qian D, Liu WK, Ruoff RS (2001) Mechanics of C60 in nanotubes. J Phys Chem B 105:10753–10758
7. Bower C, Rosen R et al (1999) Deformation of carbon nanotubes in nanotube-polymer composites. Appl Phys Lett 74:3317–3319
8. Chen XL, Liu YJ (2003) Square representative volume elements for evaluating the effective material properties of carbon nanotube-based composites. Comput Mater Sci 28:2394–2401

9. Joshi UA, Joshi P, Harsha SP, Sharma SC (2010) Evaluation of the mechanical properties of carbon nanotube based composites by finite element analysis. Int J Eng Sci Technol 2(5):1098–1107
10. Iijima AS, Ichihashi T (1993) Single-shell carbon nanotubes of 1-nm diameter. Nature (London) 363:603
11. Fidelus JD, Wiesel E, Gojny FH, Schulte K, Wagner HD (2005) Thermo-mechanical properties of randomly oriented carbon/epoxy nanocomposites. Compos Part A 36:1555–1561
12. Han Y, Elliott J (2007) Molecular dynamics simulations of the elastic properties of polymer/ carbon nanotube composites. Comput Mater Sci 39:315–323
13. Zhu R, Pan E, Roy AK (2007) Molecular dynamics study of the stress–strain behaviour of carbon-nanotube reinforced Epon 862 composites. Mater Sci Eng A 447:51–57
14. Wuite J, Adali S (2005) Deflection and stress behaviour of nanocomposite reinforced beams using a multiscale analysis. Compos Struct 71:388–396
15. Vodenitcharova T, Zhang LC (2006) Bending and local buckling of a nanocomposite beam reinforced by a single-walled carbon nanotube. Int J Solids Struct 43:3006–3024
16. Lei ZX, Liew KM, Yu JL (2003) Buckling analysis of functionally graded carbon nanotube-reinforced composite plates using the element-free kp-Ritz method. Compos Struct 98:160–168
17. Malekzadeh P, Shojaee M (2013) Buckling analysis of quadrilateral laminated plates with carbon nanotubes reinforced composite layers. Thin-Walled Struct 71:108–118
18. Mehrabadi Jafari S, Sobhani A, Khoshkhahesh V, Taherpour A (2013) Mechanical buckling of nanocomposite rectangular plate reinforced by aligned and straight single-walled carbon nanotubes. Compos Part B 43:160–168
19. Formica G, Lacarbonara W, Alessi R (2010) Vibrations of carbon nanotube-reinforced composites. J Sound Vib 329:1875–1889
20. Arani A, Maghamikia S, Mohammadimehr M, Arefmanesh A (2011) Buckling analysis of laminated composite rectangular plates reinforced by SWCNTs using analytical and finite element methods. J Mech Sci Technol 25:809–820
21. Shen HS (2011) Postbuckling of nanotube-reinforced composite cylindrical shells in thermal environment. Part I: axially-loaded shells. Compos Struct 93:2096–2108
22. Shen HS (2011) Postbuckling of nanotube-reinforced composite cylindrical shells in thermal environments. Part II: pressure-loaded shells. Compos Struct 93:2496–2503
23. Shen HS (2013) Thermal buckling and post-buckling behaviour of functionally graded carbon nanotube-reinforced composite cylindrical shells. Compos B Eng 50:239–247
24. Mittelstedt C (2008) Explicit analysis and design equations for buckling loads and minimum stiffener requirements of orthotropic and isotropic plates under compressive load braced by longitudinal stiffeners. Thin-Walled Struct 46:1409–1429
25. Kant T, Swaminathan K (2000) Analytical solutions using a higher order refined theory for the stability analysis of laminated composites and sandwich plates. Struct Eng Mech 10:337–357
26. Timoshenko S, Gere J (1961) Theory of elastic stability, 2nd edn. McGraw-Hill, New York
27. Kurkute JV, Shiyekar SM (2018) Buckling of a composite laminated plate subjected to the thermal environment using higher order shear and normal deformation theory. Int J Adv Res Eng Technol (IJARET) 9(5):220–228
28. Mayandi K, Jeyaraj P (2015) Bending, buckling and free vibration characteristics of FG-CNT-reinforced polymer composite beam under non-uniform thermal load. Proc Instit Mech Eng Part L J Mater Design Appl 229(1):1–9
29. Shiyekar SM, Lavate P (2015) Flexure of power law governed functionally graded plates using ABAQUS UMAT. Aerosp Sci Technol 46:51–59
30. Mindlin RD (1951) Influence of rotatory inertia and shear on flexural motions of isotropic elastic plates. J Appl Mech 18:31–38
31. ABAQUS User's Manual (2003) ABAQUS Version.6.10

Chapter 4
Experimental Investigation of Machining Parameters of Monel 400 in Powder Mixed EDM Process

Abhishek Kumar and T. Jagadeesha

Abstract Electric Discharge Machining (EDM) is basically a die-sinking process. First, material is removed where the current density is high, and a replica of the tool is reproduced on the work. The productivity of EDM can be increased by increasing the spark frequency. This is done by introducing conductive metal powder. Process performance of EDM depends both on the tool as well as the work material. In this current research work, the effect of graphite powder of size 30–40 μ is used with a combination of Monel 400 as a workpiece material and copper as a cathode. Monel 400 is a nickel-based superalloy material and is widely used in the chemical processing industry, automotive, marine industry, oil and gas extraction, pollution control, and waste processing industry, and high-temperature applications. Processes parameters chosen for the experiments include powder concentration, peak current and pulse-on time. The effect of these parameters on Tool Wear Rate (TWR) and Material Removal Rate (MRR) is investigated. The Taguchi orthogonal array was used as the basis for the studies. Peak current and tool material were shown to have a significantly higher impact on MRR and TWR using ANOVA. A hybrid method of dimensional and regression analysis was used using MINITAB to build semi-empirical models for material removal rate based on machining parameters and significant thermo-physical characteristics.

Keywords Powder mixed Electrical discharge machining · Monel 400 · Dimensional analysis · Regression analysis

---

A. Kumar · T. Jagadeesha (✉)
Department of Mechanical Engineering, National Institute of Technology, Calicut, Kerala 673 601, India
e-mail: jagdishsg@nitc.ac.in

V. Bindhu et al. (eds.), *Proceedings of Sixth International Conference on Inventive Material Science Applications*, Advances in Sustainability Science and Technology,
https://doi.org/10.1007/978-981-99-4189-6_4

1 Introduction

Powder-mixed Electrical Discharge Machining (EDM) is a modern machining technique that involves the use of a conductive powder mixed with a dielectric fluid to enhance the machining performance and surface quality. The use of powder in EDM can improve machining efficiency, reduce tool wear, and enhance the surface finish of the machined parts. Figure 1 shows the typical schematic diagram of the powder-mixed EDM process.

Superalloys are a class of high-performance alloys that exhibit excellent mechanical strength, corrosion resistance, and high-temperature stability. Monel 400 is a popular nickel-copper superalloy. In recent years, the application of Monel 400 in powder-mixed EDM has gained significant attention due to its superior mechanical and chemical properties. The addition of Monel 400 powder in EDM can enhance the machining performance by reducing the Tool Wear Rate (TWR), improving the surface finish, and reducing the machining time. Moreover, using Monel 400 powder can also improve the electrical and thermal conductivity of the dielectric fluid, which can further enhance the machining performance. Buckingham Pi theorem was used to develop a model for the Material Removal Rate (MRR) with the help of multiple linear regression equations.

Investigation of electrical discharge machining with powder is of special significance. Understanding the Powder Mixed EDM (PMEDM) idea thoroughly and studying the process parameters in greater detail, especially to get improved performance and the powder used in PMEDM, plays a vital role in enhancing MRR and decreasing TWR compared to the traditional EDM process [1]. Examination of Nickel-Based Superalloy Powder Mixed EDM utilizing powdered cobalt, zinc, and molybdenum Better MRR was created by cobalt and zinc, and reduced tool wear rate was obtained by molybdenum powder [2]. The potential of graphite powder is realized in increasing the machining rate in AEDM of nickel-based Super Alloy 718. The overall machining rate increases by 26.85 if fine graphite powder contributes 2.4% to the process [3].

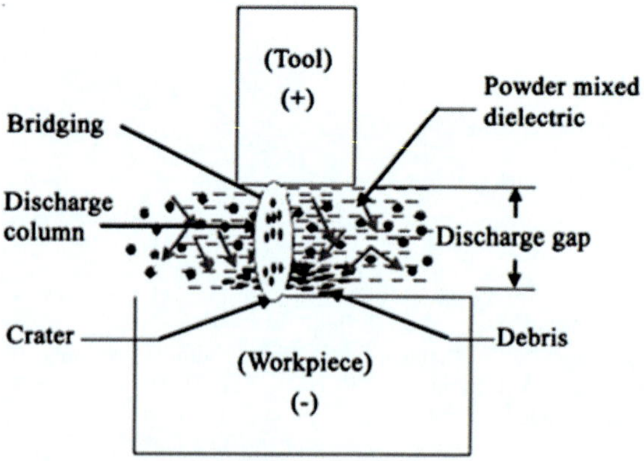

Fig. 1 Schematic diagram of the PMEDM process

Semi-empirical models have been created for the responses, using a hybrid method, for the PMEDM process of aluminum/alumina composites with metal matrix, and the PCA technique has been used to determine the weights for responses while GRA has been used to combine the multiple objectives into single [4]. The frequency of sparking and energy per spark are important parameters that affect the MRR [5]. Arching in powder EDM gives a rough surface and has to be avoided under all circumstances. It is found that silicon powder gives more arching during PMEDM and thus, surface roughness is greatly affected [6]. EDM performance has been identified as particle size, particle concentration, particle density, electrical resistivity, and thermal conductivity [7, 8].

2 Materials and Experimental Conditions

In this experiment, the graphite powder of size 30–40 μm is used with cathode as a copper rod of diameter 11 mm and height 50 mm, with 99% purity. Work material is nickel-based superalloy, Monel 400 rod of diameter 10 mm and height 30 mm. Physical properties of Monel 400 are given in Table 1. The various process parameters such as sparking voltage, pulse on–off time, and current are given in Table 2. A complete experimental setup to carry out powder based EDM is shown in Figs. 2 and 3.

Table 1 Physical properties of Monel

Parameter	Value
Density (kg/m^3)	8800
Melting temp. (°C)	1325
Thermal conductivity (W/mk)	21.8
Specific heat capacity (J/kgK)	427
Tensile strength (MPa)	550

Table 2 Process parameters for the experiment

Conditions	Value
Sparking voltage (V)	40
Peak current (I)	12, 15, 28
Pulse on time (μ_s)	20, 40, 75
Pulse off time (μ_s)	10
Powder concentration (g/l)	0, 2.5, 5
Tool polarity	Negative
Dielectric used	DI water
Tool materials	Copper
Work materials	Monel 400

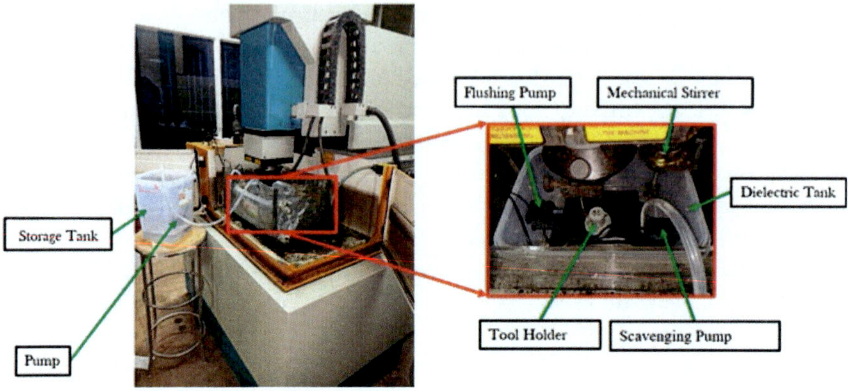

Fig. 2 Experimental setup of PMEDM

Fig. 3 Comparison of Avg. MRR

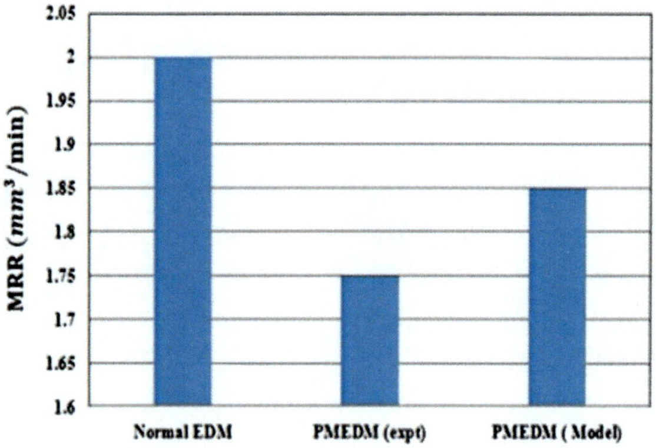

The precise weighing scale is used to calculate the weight of the tool/work before and after machining, and weight loss is calculated based on the loss of tool/work material. TWR or MRR is calculated using the theoretical formula

$$\text{MRR or TWR} = \frac{(\text{Wb} - \text{Wa}) \times 1000}{\rho t} \text{mm}^3/\text{min}$$

where

Wb = weight of the anode or cathode prior to machining (g);
Wa = weight of the anode or cathode after post-machining (g);
t = machining time for each trial (s);
ρ = density of material (g/cm^3).

3 Mathematical Modeling of MRR Using Dimensional Analysis

The mathematical relationship between MRR and process parameters is expressed using the principle of dimensional analysis and the Buckingham π theorem. The various process factors, their units, and dimensions are given in Table 3. Multiple linear regression analysis is used to calculate the MRR.

Mathematically, Buckingham π theorem is expressed as

$$\mathrm{MRR(Q)} = f\left(\mathrm{S}, \rho, \mathrm{I_p}, \mathrm{V_g}, \mathrm{T_{on}}, \mathrm{K}, \mathrm{P_c}, \mathrm{C}\right)$$

$$\mathrm{n} = 9, \mathrm{m} = 5, \mathrm{n} - \mathrm{m}, = 4$$

$$\mathrm{n} = \text{number of variables}, \mathrm{m} = \text{no.of fundamental dimensions},$$

$$\mathrm{n} - \mathrm{m} = \text{no.of } \pi \text{ terms}$$

Each π term is expressed as a function given by Eq. (1).

$$f(\pi_1, \pi_2, \pi_3, \pi_4) = 0 \tag{1}$$

$$\pi_1 = \mathrm{S^{a1}} \rho^{\mathrm{b1}} \mathrm{I_p^{c1}} \mathrm{K^{d1}} \mathrm{V_g^{e1}} \mathrm{Q}$$

$$\pi_2 = \mathrm{S^{a2}} \rho^{\mathrm{b2}} \mathrm{I_p^{c2}} \mathrm{K^{d2}} \mathrm{V_g^{e2}} \mathrm{T_{on}}$$

$$\pi_3 = \mathrm{S^{a3}} \rho^{\mathrm{b3}} \mathrm{I_p^{c3}} \mathrm{K^{d3}} \mathrm{V_g^{e3}} \mathrm{P_c}$$

$$\pi_4 = \mathrm{S^{a4}} \rho^{\mathrm{b4}} \mathrm{I_p^{c4}} \mathrm{K^{d4}} \mathrm{V_g^{e4}} \mathrm{C}$$

Table 3 Response and variables dimension

Factor	Unit	Dimension
MRR (Q)	mm^3/min	L^3T^{-1}
Powder size (S)	μm	L
Density (ρ)	g/cm^3	ML^{-3}
Peak current (I_p)	A	I
Gap voltage (V_g)	V	$ML^2T^{-3}I^{-1}$
Pulse on time (T_{on})	μs	T
Thermal conductivity (K)	W/mK	$MLT^{-3}\theta^{-1}$
Powder concentration (P_c)	g/l	ML^{-3}
Specific heat (C)	J/KgK	$L^2T^{-2}\theta^{-1}$

By solving the above set of equations, π_1, π_2, π_3 *and* π_4 are expressed as

$$\pi_1 = \left(\frac{\rho}{I_p V_g}\right)^{0.33} \frac{Q}{S^{1.33}}$$

$$\pi_2 = \left(\frac{I_p V_g}{\rho}\right)^{0.33} \left(\frac{T_{on}}{S^{1.66}}\right)$$

$$\pi_3 = \left(\frac{P_c}{\rho}\right)$$

$$\pi_4 = \frac{\left(S I_p V_g \rho^2\right)^{0.33} C}{K}$$

By expressing π_1 in terms π_2, π_3 *and* π_4, Eq. 2 is obtained.

$$\pi_1 = f(\pi_2, \pi_3, \pi_4) \tag{2}$$

Using the principle of Buckingham π theorem, π_1 in exponentials form is

$$\pi_1 = M\left(\pi_2^{\beta_1}\right)\left(\pi_3^{\beta_2}\right)\left(\pi_4^{\beta_3}\right)$$

The final form of the equation is given in Eq. 3.

$$\left(\frac{\rho}{I_p V_g}\right)^{0.33} \left(\frac{Q}{S^{1.33}}\right) = M\left[\left(\frac{I_p V_g}{\rho}\right)^{0.33}\left(\frac{T_{on}}{S^{1.66}}\right)\right]^{\beta_1}\left(\frac{P_c}{\rho}\right)^{\beta_2}\left(\frac{\left(S I_p V_g \rho^2\right)^{0.33} c}{K}\right)^{\beta_3} \tag{3}$$

Taking logarithms on both sides of Eq. 3

$$\log \pi_1 = \log M + \beta_1 \log(\pi_2) + \beta_2 \log(\pi_3) + \beta_3 \log(\pi_4)$$

Multiple linear regression equation thus becomes where

$$Y = \beta_0 + \beta_1 X1 + \beta_2 X2 + \beta_3 X3 \tag{4}$$

Y = response variable
β = regression coefficient
X = set of 'n' dependent variables.
Regression coefficient obtained after solving Eq. 4 is given as

$$\beta_0 = \log M = -31 \quad \beta_1 = 1.9$$
$$\beta_2 = 1.4 \quad \beta_3 = 10.12$$

Inserting all the regression coefficients, the MRR equation is obtained as

$$\mathrm{MRR(Q)} = \left(\frac{S^{1.33}}{\left(\frac{\rho}{I_p V_g}\right)^{0.33}} \right) M \left[\left(\frac{I_p V_g}{\rho} \right)^{0.33} \left(\frac{T_{on}}{S^{1.66}} \right) \right]^{1.9} \left(\frac{P_c}{\rho} \right)^{1.4} \left(\frac{(S I_p V_g \rho^2)^{0.33} c}{K} \right)^{10.12} \tag{5}$$

Rearranging the terms in Eq. (5)

$$\mathrm{MRR(Q)} = \frac{(S)^{1.8} (P_c)^{1.4} (\rho)^{4.5} (I_p V_g)^{4.3} (T_{on})^{1.9} (C)^{10.12}}{(K)^{10.12} \times (10)^{36}}$$

4 Results and Discussion

Twenty seven experiments with three factors and three variables are conducted experimentally. MRR of the workpiece material and tool wear of the cathode is calculated by the weight difference. The results are compared with the mathematical model developed. The experimental and mathematical modeling results are given in Table 4.

Material removal rate in normal EDM and power mixed EDM are calculated experimentally and compared with a mathematical model developed through dimensional analysis. Figure 3 shows the average values of MRR for each configuration. Tool wear rates in normal EDM and powder mixed EDM are evaluated experimentally, and the results are shown in Fig. 4.

5 Conclusion

In this research work, various experiments are carried out to find out the Material Removal Rate (MRR) and tool wear rate with graphite powder in Electrical Discharge Machining (EDM) process. The detailed mathematical model has been developed using dimensional analysis principles. It is observed that MRR in normal EDM is slightly higher than Powder mixed EDM (PMEDM) with graphite powder in the concentration range of 2.5 to 5 g/l. This lower MRR is due to the concentration range. MRR is also evaluated using a mathematical model developed using Buckingham π theorem, and it is found that mathematical model results are close to experimental value. The deviation between mathematical modeling and experimental method is approx. 6.18%. Therefore, it can be concluded that a developed mathematical model can be used to predict the MRR in PMEDM effectively with an acceptable error ranging between 6 and 10%. PMEDM graphite powder is found to give less tool

Table 4 Experimental and Predicted value of MRR and TWR

Sl no	T_{on}	I_p	P_c	MRR (Expt)	MRR (Model) PMEDM	TWR
1	20	12	0	1.138	0	0.132
2	40	12	0	2.069	0	0.066
3	75	12	0	2.425	0	0.276
4	20	15	0	1.217	0	0.125
5	40	15	0	2.191	0	0.161
6	75	15	0	2.591	0	0.020
7	20	18	0	0.898	0	0.032
8	40	18	0	2.225	0	0.116
9	75	18	0	3.419	0	0.118
10	20	12	2.5	1.633	0.058	0.097
11	40	12	2.5	1.643	0.218	0.062
12	75	12	2.5	1.694	0.72	0.053
13	20	15	2.5	0.621	0.152	0.102
14	40	15	2.5	1.802	0.569	0.055
15	75	15	2.5	2.036	1.881	0.036
16	20	18	2.5	1.465	0.334	0.095
17	40	18	2.5	2.047	1.247	0.079
18	75	18	2.5	2.584	4.119	0.034
19	20	12	5	0.353	0.154	0.146
20	40	12	5	0.910	0.576	0.115
21	75	12	5	1.334	1.901	0.107
22	20	15	5	0.482	0.402	0.137
23	40	15	5	1.532	1.503	0.119
24	75	15	5	3.403	4.964	0.131
25	20	18	5	2.418	0.882	0.14
26	40	18	5	2.915	3.293	0.105
27	75	18	5	3.005	10.872	0.045

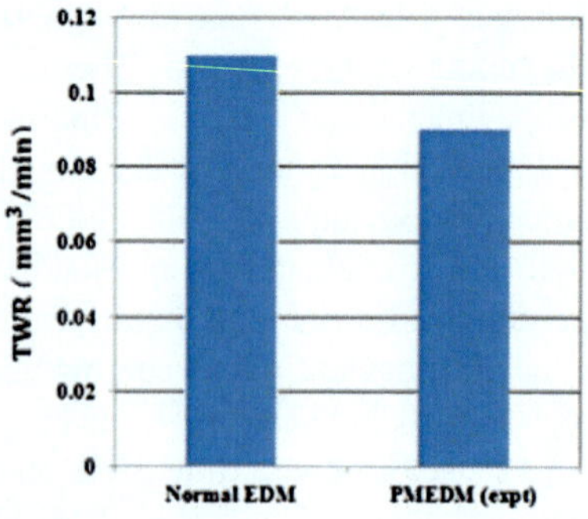

Fig. 4 Comparison of Avg. TWR

wear rate compared to normal EDM, and therefore, PMEDM is very effective in increasing the productivity by increasing the tool life. Reduction of tool wear up to 21.85% is obtained in experimental PMEDM.

References

1. Ayanesh Joshi Y, Anand Joshi Y (2019) A systematic review on powder mixed electrical discharge machining. Heliyon 5(12):1–12
2. Ramesh S, Jenarthanan MP (2021) Investigation of powder mixed EDM of nickel-based superalloy using cobalt, zinc and molybdenum powders. Int J Adv Manuf Technol 74:923–936
3. Kumar A, Maheshwari S, Sharma C, Beri N (2010) Realizing potential of graphite powder in enhancing machining rate in AEDM of nickel based super alloy 718. Proc Int Conf Adv Mech Eng 1:1–4
4. Talla G, Gangopadhyay S, Sahoo DK, Biswas CK (2015) Modeling and multi-objective optimization of powder mixed electric discharge machining process of aluminum/alumina metal matrix composite. Eng Sci Technol Int J 18:369–373
5. Bhattacharya A, Batish A, Singh G (2012) Optimal parameter settings for rough and finish machining of die steels in powder-mixed EDM. Int J adv Manuf Technol 61:537–548
6. Pecas P, Henriques E (2003) Influence of silicon powder-mixed dielectric on conventional electrical discharge machining. Int J Mach Tools Manuf 43:1465–1471
7. Tzeng Y-F, Lee C-Y (2001) Effects of powder characteristics on electrodischarge machining efficiency. Int J Adv Manuf Technol 17:586–592
8. Chakraborty S, Dey V, Ghosh SK (2015) A review on the use of dielectric fluids and their effects in electrical discharge machining characteristics. Precis Eng 40:1–6

Chapter 5
Synthesis of Estolides from Coconut, Sunflower and Karanja Oils

P. V. Maneesh Kumar and T. Jagadeesha

Abstract Lubricating machines with mineral-based lubricants have become a major cause of pollution now. To use natural oils for making lubricants instead of petroleum is something mainstream researchers have been looking at for a few decades. Though vegetable oils show excellent biodegradability, their cold flow properties are poor compared to functional lubricants. Biodegradable lubricant oils must have excellent cold flow properties as well as competitive costs in order to be accepted widely as usable base stock to industrially usable lubricants. Cold flow characteristics of natural oils are generally subpar to industrial lubricants. This is one major drawback that limits the usage of otherwise abundant vegetable oil. Though vegetable oils show excellent biodegradability, their cold flow properties are poor compared to functional lubricants. In this present study, three vegetable oils have been synthesized to obtain their estolides, they are coconut, sunflower and karanja oils, to analyze the variations. SPARTAN molecular dynamics software is utilized in parallel to understand the molecular structure of products.

Keywords Estolides · SPARTAN · Saturated fatty acids · Pour point

1 Introduction

Lubrication is one essential aspect of our industrial civilization. It has been practiced for thousands of years. Majority of lubricants consumed today are derived from mineral oils and it has become one major environmental concern. Estolides belong to the family of esters synthesis from natural oils as the carboxylic acid functional group of the fatty acid forms an ester linkage with the unsaturated site of another fatty acid [1]. Estolide number represents the mean number of additional fatty acids attached to base fatty acids. This value changes with respect to temperature, type of

P. V. Maneesh Kumar (✉) · T. Jagadeesha
Department of Mechanical Engineering, National Institute of Technology, Calicut, Kerala 673 601, India
e-mail: maneesh_p210081me@nitc.ac.in

V. Bindhu et al. (eds.), *Proceedings of Sixth International Conference on Inventive Material Science Applications*, Advances in Sustainability Science and Technology,
https://doi.org/10.1007/978-981-99-4189-6_5

saturated fatty acid being supplied and other various parameters. Lubricants of today are 90% base stock with 10% additives [2]. Continuous use of these mineral-based lubricants leads to environmental issues as well as health problems for its operators. Thus, alternative lubricants synthesized from vegetable oils are Generally Regarded as Safe products. Vegetable oils are an eco-friendly alternative for the base stock of industrial lubricants as they are more biodegradable and less toxic. Vegetable oils have many positive properties of lubricants, such as high viscosity index, improved contact lubrication, good flash points and very low volatility. Vegetable oils show a wide range of viscosities, whereas mineral oils have an average viscosity range of 441 mPa.S at 60 °C [3]. However, vegetable oils possess some weaknesses. One major weakness of vegetable oils is their subpar cold-flow properties. At lower temperatures, vegetable oils lose their fluidity due to clouding, thus, bad cold-flow properties of natural oils are one important challenge for using them as lubricants or as base stock of lubricants.

Estolides are esters derived by modifying fatty acids. They have found application in many cosmetics, bio-lubricants, and coatings. Estolide synthesis occurs when the fatty acids stack on the unsaturated sites of other fatty acids. Estolide number is the mean number of additional fatty acids branched on the base. These ester linkages are strong against hydrolysis, giving them improved properties against pure natural oils [4]. Estolides are naturally occurring and are also manufactured in labs [5]. We have synthesized oleic estolides further reacted to 2-ethylhexanol.

$$\text{Oleic acid} + \text{Saturated fatty acid} + HClO_4 = \text{Free Estolide}$$

$$\text{Free Estolide} + \text{2-ethylhexanol} = \text{Estolide 2-ethylhexyl ester}$$

SPARTAN is a molecular dynamics software working on quantum chemical calculations to formulate the structure of molecules. The structure alone will be providing valuable insights into the properties and the variations caused by the molecular structure. The objective of the present study is to synthesize estolides from these three natural oils and to compare their properties over their base oil to analyze the improvements along with molecular dynamics analysis.

2 Experimental Procedure

2.1 Materials

Coconut, Sunflower, Karanja and Castor oils were procured locally as they are readily available for cooking purposes. Oleic acid, Perchloric acid, 2-ethyl hexanol, KOH, Ethanol and Silica gel was obtained from SIGMA ALDRICH Chemical Company, supplied by Alpha Chemicals, Ernakulam.

2.2 Preparation of Estolides

Estolides were synthesized from Coconut, Sunflower and Karanja oils through condensation reactions catalyzed with acids. The reactions took place in a 750 ml 3-necked flask. The reaction was conducted at standard atmospheric conditions. The reactant mixture for estolide contains 90 g base oil, 10 g of castor oil and 20.3 g of oleic acid. The mixture was heated in a Rota mantle to a temperature of 60 ± 0.1 °C. A magnetic stirrer was used for mixing. Once the temperature reached 60 °C 3 g of Perchloric acid ($HClO_4$) was incorporated drop by drop. Then the reactants were heated for 24 h. After 24 h, 56.5 g of 2-ethyl hexyl alcohol was added to the reactant flask and the reactants were further heated for 2 h. For quenching, 2 g potassium hydroxide (KOH) mixed through ultrasonic stirring for 180 s in a 90% ethanol–water mixture was added. The synthesis of estolide ester in Rota Mantle is shown in Fig. 1. Water washing in the separating funnel is shown in Fig. 2.

The completed reactions were cooled for 30 min. The reactant mixture was moved to a separating funnel, and 100 g hexane was added for extracting the organic layer. The estolide synthesized was then further water-washed with warm water at 60 °C till the pH of washed water obtained equals the pH of normal water. For further water removal, the estolides were kept in an open beaker surrounded by silica gel in a closed vessel for 24 h. Commercially available coconut, sunflower and karanja oils were used for experiments. Estolide ester placed between silica gel for water absorption is shown in Fig. 3. Coconut, sunflower and karanja oils and estolide are shown in Figs. 4 and 5.

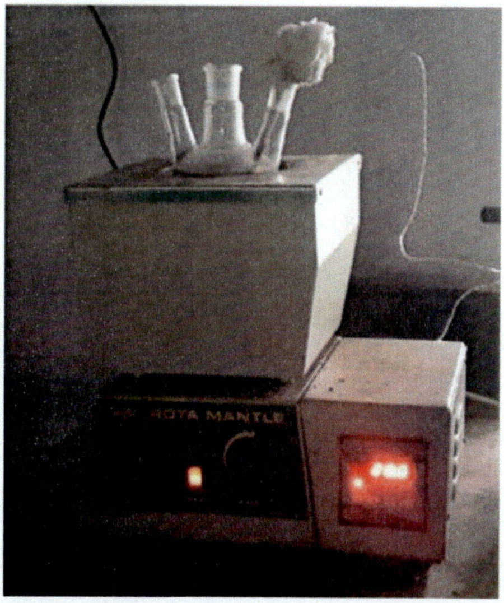

Fig. 1 Synthesis of estolide ester in Rota Mantle

Fig. 2 Water washing in separating funnel

Fig. 3 Estolide ester placed between silica gel for water absorption

Fig. 4 Coconut, sunflower and Karanja oils

Fig. 5 Coconut, sunflower and Karanja estolides

2.3 *Pour-Point Analysis*

Pour point of the oils and synthesized estolides were evaluated in a pour-point testing apparatus with iso-propyl alcohol coolant as per ASTM D 97 standards. The oils were placed in apparatus in a test tube with a thermometer stoppered on it at room temperature and were kept for cooling. At every 3 °C reduction in temperature, the test tube was taken out to check fluidity by tilting. Cloud point, which is the temperature at which clouds start to form, is analyzed visually and is noted. Pour points are also evaluated visually and are noted.

2.4 Quantum Chemical Analysis

SPARTAN is a molecular dynamics software that works on the basis of quantum chemical applications that can give the appropriate molecular structures of complex chemicals. The estolide esters synthesized were given to SPARTAN for understanding their chemical structure and to estimate their electrostatic potential.

3 Results and Discussion

3.1 Quantum Chemical Analysis

Initially, base fatty acids were imagined to analyze and predict the tribological properties. Since estolide formation is prone to fatty acids with double bonds in them, saturated fatty acids were analyzed along with their estolides to compare. We can observe estolides as branched structures with loosely distributed Electrostatic potentials. This variation with the regular base oil configuration will surely cause variations in the properties as crystallization is attained by the uniform stacking of molecules over one another and the variation in the structures, especially from linear-easily stackable structures to branched—not so easily stackable structures will show the variations.

From the structure of these molecules (Figs. 6, 7, 8, 9 and 10), it is evident that the added molecular mass will improve the viscosity of the liquid. The complex structure of the molecules will make the crystallization a slow process delaying the pour points.

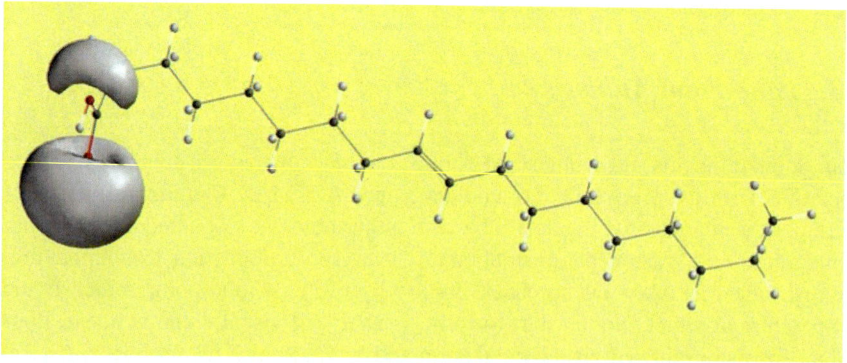

Fig. 6 Molecular structure of oleic acid shown with electrostatic potential (grey cloud)

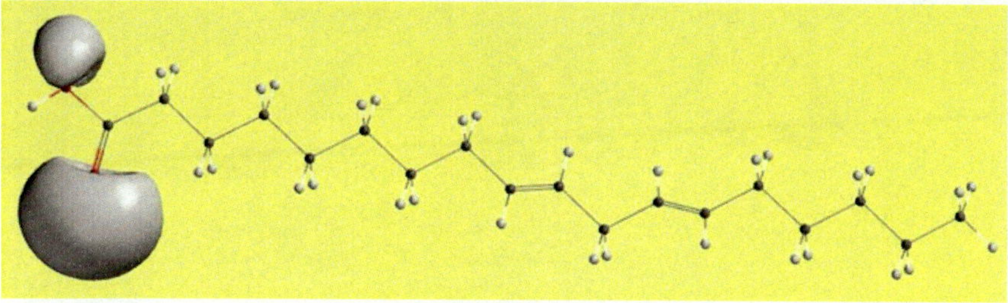

Fig. 7 Molecular structure of linoleic acid shown with electrostatic potential

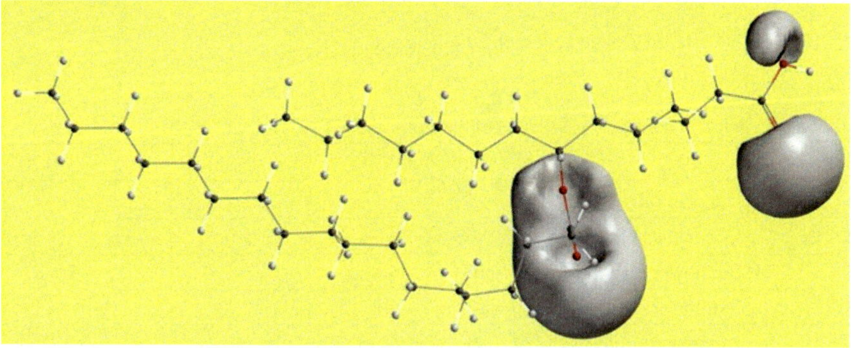

Fig. 8 Estolide of 2 oleic acid molecules (N = 1)

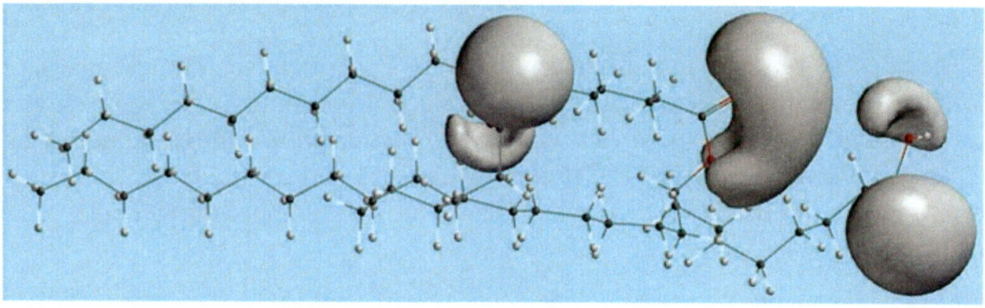

Fig. 9 Estolide of 2 oleic acid with a linoleic acid (N = 2)

3.2 *Pour-Point Analysis*

Pour point is one major aspect of lubricants and is one area where natural oils show poor properties. Cloud points of base oils and estolides were evaluated using ASTM D 97 standards. Variation in cold flow properties of base oils and estolides SAE20W40 is shown in Table 1.

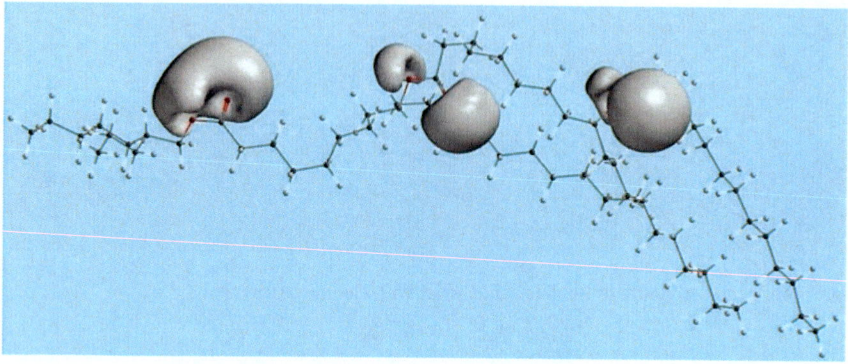

Fig. 10 Estolide ester of 2 oleic acid with a linoleic acid molecule

Table 1 Variation in cold flow properties of base oils and estolides SAE20W40

Sl no	Sample	Cloud point °C	Pour point °C	ΔT
1	Coconut oil	25	22	−24
2	Coconut estolide	0	−2	
3	Sunflower oil	−16	−18	−3
4	Sunflower estolide	−18	−21	
5	Karanja oil	6	9	−21
6	Karanja estolide	−6	−12	
7	SAE20W40	−18	−21	

The samples were kept in a test tube with a −50 °C thermometer. The test tube was placed in a cooling apparatus with isopropyl alcohol for cooling. The temperature at the beginning of the experiment was at room temperature and the reduction in temperature was noted at every 3 °C to check if clouding had occurred. Once clouding begins clotting of oils is to follow, which shows the pour point of the sample.

4 Conclusion

Vegetables stack on each other as temperature reduces on the path to crystallization, as branches form on these structures, stacking is delayed leading to a reduced pour point [6]. Pour points, which are the major demerits of vegetable oils, can be improved by attaching other fatty acid chains of similar or other oil molecules with appropriate chemical processes. This hinders the close packing of these molecules while solidification occurs. Their cold-flow properties show significant improvements. Unsaturation/Double bond sites present in mono- and polyunsaturated fatty acids can be suitable locations for attaching branches. The reduction of internal symmetry associated with branching also slows down the stacking, reducing the pour-point further in the temperature scale [7]. Further addition of 2-ethylhexanol for capping increases the molecular weight and forms an additional branch [8]. As estolides are obtained from vegetable oils, it is normal to expect them to have poor cold-flow properties, but the results are showing very good improvement in this area to a point that is even comparable with commercial lubricants. We can source raw materials required for most of our applications from the plants itself [9]. Recent developments in this area are reassuring the benefits of plant-based industrial fluids [10]. Polyols obtained from natural oils are another well-behaved bio-based lubricant capable of huge improvements [11]. Another positive information is that the characteristics can be further improved by varying the molecular weight of components [12]. The capability of estolide esters to become biofuel is another strength of this bio-based molecule [13]. The biofuel thus created is also showing very good cold flow characteristics [14]. and there are many novel methods for further advancing the cold flow capabilities of these liquids [15]. Industrial fluids are being replaced by natural oil based fluids and are visible in many industries [16]. Additionally, nanomaterials can be used for improving these fluids, and this area is to be further investigated.

References

1. Cermak SC, Bredsguard JW, Dunn RO, Thompson T, Feken KA, Roth KL, Kenar JA, Isbell TA, Murray RE (2014) Comparative Assay of Antioxidant Packages for Dimer of Estolide Esters. J Am Oil Chem Soc
2. Ho CK, McAuley KB, Peppley BA (2019) Biolubricants through renewable hydrocarbons: a perspective for new opportunities. Renew Sustain Energy Rev
3. Bin X, Ding R, Yang Z, Sun Y, Zhang J, Kaiji L, Cao D, Gao A (2023) Investigation on performance of mineral-oil-based rejuvenating agent for aged high viscosity modified asphalt of porous asphalt pavement. J Clean Prod 395:136285
4. Garcia-Zapateiro LA, Franco JM, Valencia C, Delgado MA, Gallegos C (2013) Viscous, thermal and tribological characterization of oleic and ricinoleic acids-derived estolides and their blends with vegetable oils. J Indus Eng Chem
5. Cermak SC (2001) Synthesis of estolides from oleic and saturated fatty acids. J Am Oil Chem Soc 06
6. Ajithkumar G (2009) Analysis of the pour point of coconut oil as a lubricant base stock using differential scanning calorimetry. Lubric Sci 01
7. Greco-Duarte J, Collaço ACA, Costa AMM, Silva LO, Da Silva JAC, Torres AG, Fernandez-Lafuente R, Freire DMG (2019) Understanding the degree of estolide enzymatic polymerization and the effects on its lubricant properties. Fuel
8. Isbell TA, Edgcomb MR, Lowery BA (2001) Physical properties of estolides and their ester derivatives. Indus Crops Prod
9. Ray AK, Stark RE (1998) Isolation and molecular structure of an oligomer produced enzymatically from the cuticle of lime fruit. Phytochemistry 48(8):1313–1320
10. Perera M, Yan J, Xu L, Yang M, Yan Y (2022) Bioprocess development for biolubricant production using non-edible oils, agro-industrial byproducts and wastes. J Clean Prod 357:131956

11. Åkermana CO, Gaber Y, Abd Ghani N, Lämsä M, Hatti-Kaul R (2011) Clean synthesis of biolubricants for low temperature applications using heterogeneous catalysts. J Mol Catal B Enzym 72:263–269
12. Bantchev GB, Cermak SC (2022) Correlating viscosity of 2-ethylhexyl oleic estolide esters to their molecular weight. Fuel 309:122190
13. de Haro JC, del Prado Garrido M, Perez A, Carmona M, Rodríguez JF (2018) Full conversion of oleic acid to estolides esters, biodiesel and choline carboxylates in three easy steps. J Clean Prod 184:579–585
14. Smith PC, Ngothai Y, Nguyen QD, O'Neill BK (2010) Improving the low-temperature properties of biodiesel: methods and consequences. Renew Energy 35:1145–1151
15. Sierra-Cantora JF, Guerrero-Fajardo CA (2017) Methods for improving the cold flow properties of biodiesel with high saturated fatty acids content: a review. Renew Sustain Energy Rev 72:774–790
16. Zhang C, Garrison TF, Madbouly SA, Kessler MR (2017) Recent advances in vegetable oil-based polymers and their composites. Prog Polym Sci 71:91–143

Chapter 6
Innovative Methods for Increasing the Tool Life of Cutting Tools in Complex Surface Treatment

J. O. Sharipov

Abstract In this article, a high-speed cutting tool with complex surface treatment: ion nitriding, ion alloying and coating was used to cut difficult-to-cut chromium-nickel material. The effectiveness of chemical-thermal treatment before coating and its effect on the physicochemical properties has been studied. In the article, for the first time, the process of ion nitriding was carried out to harden the base casting, which contributes to surface hardening at a thickness of 80 microns. After the ion doping process, the surface is coated with (TiAl)N. The processed cutting tools were tested at the machine-building plant of the metallurgical plant. Complex processing of the cutting tool was carried out on a Fus-32 horizontal milling machine in the machine-building shop of the metallurgical plant. This article uses innovative methods of ion nitriding, ion alloying and coating in the complex processing of cutting tools. The main task of the research is to reduce the plastic deformation of the coating on the surface of the cutting tools working under high load.

Keywords High-speed tool · Nitriding · Alloying · Covering · Wear resistance · Hard alloy

1 Introduction

In spite of the year-after-year increasing application of tools made of hard alloys, cutting ceramics and extra-hard materials, the number of high-speed steels used in the manufacture of metal-working tools does not decrease at all. In our study of heat treatment and coating, we are dealing with work hardening caused by the passage of an elastic wave generated by a precisely pulsed electron beam. However, due to the short duration of the process and thermal inertia, the heating caused by compression and internal friction, most likely, should not be a physical factor that determines the

J. O. Sharipov (✉)
Bukhara Engineering-Technological Institute, Bukhara, Uzbekistan
e-mail: jamshid_sharipov_85@mail.ru

V. Bindhu et al. (eds.), *Proceedings of Sixth International Conference on Inventive Material Science Applications*, Advances in Sustainability Science and Technology,
https://doi.org/10.1007/978-981-99-4189-6_6

behavior of the material under such conditions. In this case, the key role should be played by the mechanical activation of high-speed physicochemical processes that inevitably occur in both liquid and solid phases. The appearance of a melt causes a sharp increase in the interfacial surface and an increase in the rate of the nitride formation reaction due to the additional release of energy during the exothermic reaction. At present, high-speed steel tools with various durable coverings based on nitrides of refractory metals obtained by the method of physical deposition are widely used [1].

Today, in order to create a high-speed cutting layer tool that slows down the rate of its erosion and has characteristics that meet many requirements, as well as to determine the conditions for the rational use of complex cutting tools research is being carried out and optimization of combined processing technologies remains relevant.

Covering contributes to a significant increase in the performance of tools made from these alloys at high cutting speeds. However, the efficiency of such tools decreases sharply as the increases the thickness of the shaving being cut. Under these conditions, carbide tools with a higher cobalt content work much better, as they have high bending strength and toughness, as well as a reduced tendency to brittle fracture.

A cutting tool intended for production is very important, and in some cases decisive, for the correct surface finish of the material. The reason for the low margin of stability of the cutting tool is its rapid wear or failure of the working surface. There are several ways to increase its stability, of which coating is the most common at present. These coatings are applied to the surface of the cutting tool and parts in two ways: physical and chemical.

In this research work, the difficult-to-machine heat-resistant chromium-nickel alloy 34XH1MA was processed on a Fus-32 horizontal milling machine with a cutting tool made of R6M5 high-speed tool steel [2].

High-speed steel tools have great advantages at low and medium cutting speeds in combination with medium and large shear sections, as they have relatively low heat resistance, medium hardness and the highest bending strength and toughness, as well as a high endurance limit. These same tools operate better in conditions that take advantage of the small radius of rounding that is characteristic of tools made of high-speed steels.

This is due to the fact that parts with certain requirements for the shape and quality of the surface in mechanical engineering are processed by metal cutting, since the implementation of such technical requirements for other processing methods is difficult. In machining, problems arise due to a number of production requirements on the part of designers and technologists: frequent failure or erosion of the cutting tool, improperly selected cutting mode, the use of a coating that is not intended for the material being processed, when using a cutting tool, the output.

The quality of manufactured parts, the level of accuracy, the productivity and efficiency of the machining process and the conditions of automated production and adaptive production systems mainly depend on the quality, reliability and performance of cutting tools used in mechanical engineering.

From the review of the literature, it was found that the main conditions for the use of cutting tools are based on the creation of a nitrogenized, i.e., reinforced layer, which develops one or another structural composition with a certain thickness and microhardness, which determines the performance of high-speed steel during the cutting process.

2 Materials and Methods

At present, the method of ion nitriding is rapidly developing and is widely used in the production of machine-building industry. Ion nitriding technology is considered to be simple and economical compared to carburizing and nitrocement technologies and is one of the last stages of parts manufacturing. Before placing the workpiece in the working chamber, it is degreased in gasoline until the conditional time of ion-nitriding.

Specially designed cutting disk mills with various options for combined surface ion-plasma processing were used as a cutting tool for milling (Fig. 1). The process of ion nitriding of a cutting tool made of R6M5 high-speed tool steel was carried out on an APP-2 unit. Thermal treatment was carried out at a temperature of 470 °C for 45 min, which makes it possible to create a thermochemically hardened layer up to 40 mm thick and with microhardness up to $HV_{50} = 140$ MPa on the surface. The ion nitriding process makes it possible to saturate cast iron, carbon, alloyed, tool, corrosion-resistant and heat-resistant steels, titanium and cermet hard alloys with nitrogen. The final wear-resistant covering (TiAl) N, (TiAl) N + ion-nitriding, (TiAl) N + ion-nitriding + ion-alloying (NbHf), with microhardness $HV_{50} = 350$ MPa was applied on a Platit π311 unit. This covering is a combination of the adhesive layer of the composition and the gradient covering (TiAl)N. In the research work, after ion nitriding and ion alloying, disk cutters were applied to a Platit p311 device.

In addition, the layer formed under the coating has a high resistance to microplastic deformation, combining hardness with extreme heat resistance. This leads to significantly reducing the power of the main heat source for the cutting edge of the tool. Ion doping of materials, or, in other words, the introduction of ions and ion implantation, is currently becoming the main technological process used to modify the electrophysical, chemical, optical, mechanical, and other properties of the surface layers of materials.

Multiphase coating layer with a grain size of 50–150 nm elements (TiAl)N, (TiAl)N + ion-nitriding, (TiAl)N + ion-nitriding + ion-doping (NbHf) up to 5 nm, on the border of which there is an amorphous the Si3N4 phase suppresses the coagulation of the grains of the main phase both during the coating process and during the operation of the cutting tool. Interfacial boundaries, which are zones of intense energy dissipation, deflect emerging cracks from the direction of propagation, partially or completely slowing them down. Before applying the wear-resistant covering, some of the samples were subjected to near-surface alloying. Processing was carried out in the RITM-SP-M installation, which is a combination of a source of low-energy

Fig. 1 High-speed R6M5 disk mills

high-current electron beams (LHEB) "RHYTHM" and two systems of magnetron sputtering on one vacuum chamber [3]. The device allows covering films on the surface of the desired product, followed by liquid-phase mixing of the film and substrate materials using an intense pulsed electron beam.

Ion doping of a cutting tool, or, in other words, the introduction of ions into its surface and the transfer of ions, is currently one of the main technological processes used to change the electrophysical, chemical, optical, mechanical and other properties of the surface layers of materials [4, 6].

Today, for a more successful implementation of complex surface treatment of high-speed tool steel, the surface of a cutting tool due to insufficient quality of preparation (quality of chemical–thermal treatment, passivity in friction, surface roughness, the presence of chips, load mode, etc.), significant wear of cutting blades and an uncontrolled amount gases and metal ions accelerated on the working surface, thickness and phase of the nitrided layer.

A multiphase structure using ion-doping in an exothermic chemical reaction between the coating metal and the substrate was obtained by depositing a thin layer of nitride-forming elements ($Nb_{72}Hf_{28}$ alloy targets were used) on the surface of the tool before treating it with a precise electron beam. Unstable iron nitrides of nitrided high-speed steel were used as its donors. The outer layer is enriched with refractory nitride phases, which, due to the extremely high cooling rate, remain small and evenly distributed in the final product. The depth of the surface layer, in which the modified steel structure is obtained, is 2–10 mm, depending on the alloying composition.

Ion-doping is based on the controlled introduction of ionized atoms and molecules into a material (solid body) accelerated in an electrostatic field. The ion-doping method is especially promising for semiconductor electronics. This method has a number of advantages over others: diffusion, melting, fusion from solution, epitaxial formation. Complex processing significantly reduces the formation of eroded grooves on the front surface of the cutting tool.

Ion-nitriding, ion-alloying processes of cutting tools are chemical–thermal treatment. Chemical–thermal treatment leads to changes in the chemical composition, structure and properties of the surface. Increasing the strength of the coating applied to the surface of the cutting tool cannot achieve the expected result from the plastic deformation of the base during processing. The role of the ion-nitrided and ion-alloyed layer in the complex processing of high-speed tool steel cutting tools is extremely large, because they ensure the strong adhesion of the coating, the reduction of plastic deformation, which is the most important condition for the use of wear-resistant cutting tools.

The phase composition and structure of the ion-nitrided layer were studied using X-ray diffraction analysis, which was carried out on an automated Dron-4 diffractometer, which was controlled by software and recorded spectra. Symmetrical reflection recording of the samples was carried out using X-ray tubes with copper and cobalt radiation.

3 Results and Discussions

Figure 2 shows the microstructure of a disk cutter with surface ion-alloyed high-speed tool steel P6M5. After each subsequent treatment, the cutting tool was checked for changes under a microscope. After milling, the change of the geometry of the cutting part of the milling cutter was carried out by measuring the radius r of the rounding of the cutting edges and checking the stability of the edges with a metallographic microscope, and then compared with the size and appearance of the initial radius, which is one of the most important factors of the cutting process.

In the process of cutting, the forces of resistance to plastic deformation of this metal act on the blade of the cutting tool. The amount of total force applied to the cutting tool to perform the cutting process must be sufficient to overcome the total resistance force of this metal (Fig. 3).

In the cutting tool we developed, material samples were exposed to a series of pulses of a low-energy high-current electron beam with a wide aperture to initiate exothermic chemical reactions between the metal coating and nitrogen, carbon or aluminum in liquid and solid phases. This paper describes the processes of ion alloying on the surface of materials using the reaction of the formation of niobium-hafnium carbide in the form of a complex carbide on the surface of the carbide phase of a hard alloy [7, 9].

As a new method for increasing the productivity of a cutting tool, the principles of a complexly machined surface (CMS) were formulated, and a special technology for implementing the CMS method was developed. The anti-corrosion complex formed during the implementation of the developed CMS technology has multi-purpose functions (Figs. 3, 4). Such a complex treatment increases the strength of the base poured under the covering, increases its adhesive properties, resistance to erosion. The covering in this complex has the property of rapid heat release in the treatment zone [10, 12]. In the process of cutting, the cutting tool blade is affected by the

Fig. 2 The nitriding process of a disk cutter in the APP-2 ion-nitriding device of cutting tools

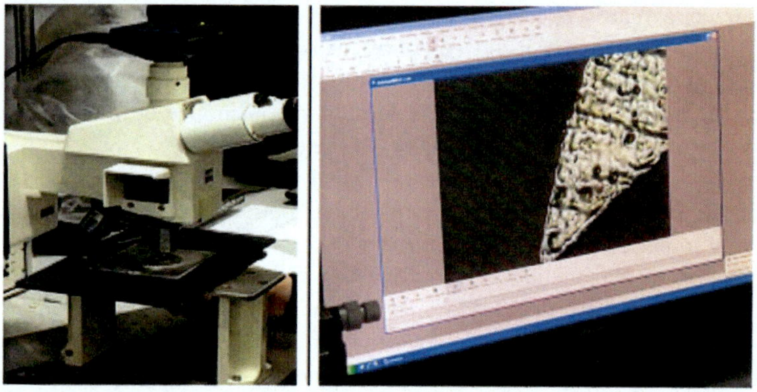

a b

Fig. 3 Microscope SteREO Discovery V12 surface structure of the cutting tool. **a** The structure of the surface of nitrided R6M5 high-speed steel after exposure to LHEB, **b** The same after exactly electron-beam alloying with zirconium

forces of resistance to the plastic deformation of this metal. The magnitude of the total force applied to the cutting tool to perform the cutting process must be sufficient to overcome the total resisting force of that metal. Several scientists have conducted research on the calculation of shear strength. It is known that [58; 18-58-b] tangential

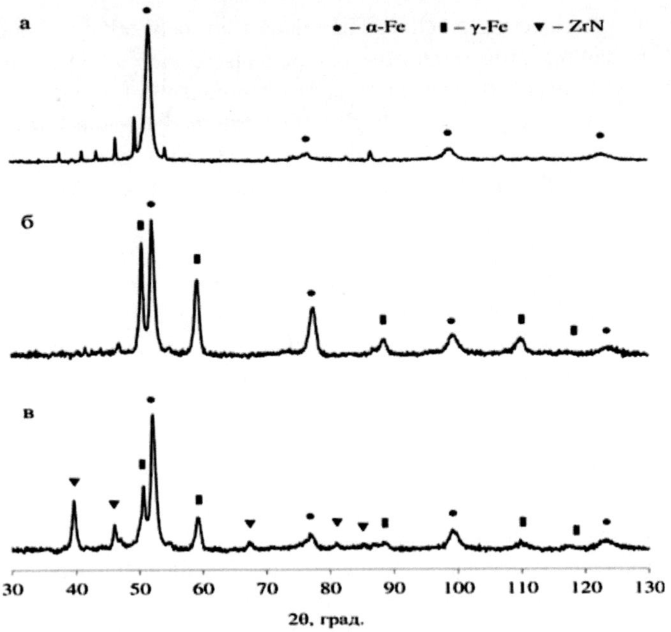

Fig. 4 Diffraction pattern of the sample surface. **a** Diffraction pattern (CoKα) from the surface of a sample of nitrided R6M5 steel, **b** after LHEB exposure, **c** after covering with NbHf film before irradiation

and radial components of shear force (P_z and P_y, respectively) were calculated for each of the indicated shear modes.

In order to determine the power spent on cutting, when calculating the strength and uniformity of the cutter, some details and parts of the machine, the cutting force R is placed on the three coordinate axes X, Y, Z and the elements P_z, P_y, P_x are formed. X is oriented along the vertical axis, and Y and Z axes are perpendicular and parallel to the detailed axis in the horizontal plane. P_z force is the main component of tangential, rotational force or shear force, P_x force is called radial force, P_y axial force or thrust force.

The magnitude of the equivalent shear force R can be determined from the following expression:

$$R = \sqrt{P_z + P_y + P_x}$$

The power consumed by the machine's prime mover electric motor is greater than the effective power consumed by the amount of losses occurring in the transmission from the electric motor to the spindle. These losses are determined by the efficiency of the transmission mechanisms.

Theoretically derived formulas are intended only for qualitative analysis of the dependence of shear force on mode forces, geometric and physical parameters, and are not used in engineering practice. The relationship between the changes in the factors of the cutting process and the components of the cutting force is determined experimentally. Instruments that measure components of shear force are called dynamometers. According to the principles of operation, dynamometers are divided into 3: there are belt-mechanical, hydraulic and belt-electric types. Cutting forces are automatically controlled by all parameters in machines controlled by a new modern digital program.

The main factors affecting the cutting forces are the physical and mechanical properties of the processed material, the geometrical parameters of the cutting tool, the cutting mode and the lubrication-cooling technological environment. For a group of materials with the same chemical composition, the increase in shear stress with increasing reinforcement far outstrips the decrease in slag penetration coefficient. Therefore, when the tensile strength increases or the Vickers hardness increases, the shear strength components increase.

The change of cutting depth t and thrust S affects the change of all three constituents of cutting force P_z, P_y, P_x. The greater the depth and thrust, the greater the cross-sectional area of the cutting layer and the volume of metal being deformed, the cutting force and the resistance to the formation of slag. However, the effect of depth of cut and thrust on cutting forces is different. A large number of experiments show that the depth of cut has a stronger effect on the components of the cutting force than thrust

$$P_{zi\ \text{чик}} = C_{pz} B^{Xpz} (S_z \sin\psi_i)^{Ypz} S_z^{-Ypz} V^{Zpz} K_{pz} \tag{1}$$

$$P_{yi\ \text{чик}} = C_{py} B^{Xpy} (S_z \sin\psi_i)^{Ypy} S_z^{-Ypy} V^{Zpy} K_{py} \tag{2}$$

$$P_{xi\ \text{чик}} = C_{px} B^{px} (S_z \sin\psi_i)^{Ypx} S_z^{-Ypx} V^{Zpx} K_{px} \tag{3}$$

The formation of chips at the beginning of the cutting process is determined by all deformation components in the field of friction and deformation of the contact surfaces when the tip of the blade cuts with less preparation (Sz) for the working stroke of the cutting tool.

When conducting research, it was found that the loads on the electric motor in the idle state and in the working state of the milling machine, when processing with milling cutters with different coatings and without coating, are different when compared. Compared to the uncoated cutters, it was found that the load on the electric motor is lower when the cutters with the complex treatment are measured in scissor ammeters.

Industrial surface modification technology is used to obtain specific properties of coatings applied in a vacuum arc discharge. Often, during coating, a negative potential is applied to the substrate, as a result of which the coating to be coated is bombarded with inert gas ions and spraying agents. In the process of cutting, the forces of resistance to plastic deformation of this metal act on the blade of the cutting tool. The amount of total force applied to the cutting tool to perform the cutting process must be sufficient to overcome the total resistance force of this metal.

Izvestno, tak kak vykhod iz stroya rejushchego instrumenta Usually, it is necessary to detail the technological equipment to understand that the technological system is full, and the instrument is not replaced. The graph of cutting tool wear versus cutting path is shown in Fig. 7.

Durability tests were carried out at IM "Navoi mechanization plant" of Navoi mining and smelting complex, when turning a forged, hard-to-cut alloy 34KhN1MA at cutting speed V = 20 m/min, feed s = 0.150 mm/rev, depth of cut t = 4 mm. The wear rate of the rear and front surfaces of 0.4 mm was chosen as the failure criterion for the cutting tool. The results obtained are presented in the Tables 1 and 2. A forged ingot of grade 34KhN1MA with a hardness of HB 375 was processed on a FUS-32 horizontal milling machine [13, 14].

The research work shows that the main factors affecting the cutting forces are the physical and mechanical properties of the material being processed, the geometric parameters of the cutting tool, the cutting mode and the lubricating and cooling technological environment [2]. When each pretty material is measured with an ambermeter, when measuring the load on the motor, a small load was found in the complex machined disc cutter (Fig. 5). The surface is strengthened during processing. The erosion of a tool is the main and important indicator of its working property. But the physical nature of deception comes from the extreme and repetitive processes of the cutting tool [15, 16] (Table 3).

When cutting with a raw tool, the characteristic point of wear was the tip of the disk mill. In the process of cutting hard-to-cut material with a complex cutting tool, it has been shown that the load on the electric motor is small, and the cutting force

Table 1 Chemical composition of the cutting tool tested preparation

Relevance of the brand according to the chemical composition				34XH1MA, chemical composition in %	
Element name	%	Element name	%	Element name	%
Carbon C	0.30	Sulfur S	0.015	Copper Cu	0.14
Silicon Si	0.34	Chrome Cr	1.40	Titan Ti	0.004
Manganese Mn	0.52	Molubden Mo	0.27	**Hardness HB 375**	
Phosphorus P	0.022	Nickel Ni	1.44		

Table 2 Physical composition of the cutting tool tested preparation

T	E 10^{-5}	α 10^{6}	Λ	P	C	R 10^{9}
Degree	Mpa	1/Deg	Vt/(м-deg)	Kg/м^3	Dj/(kg-deg)	Ом-м
20	0,74			2850		
100		23,2				

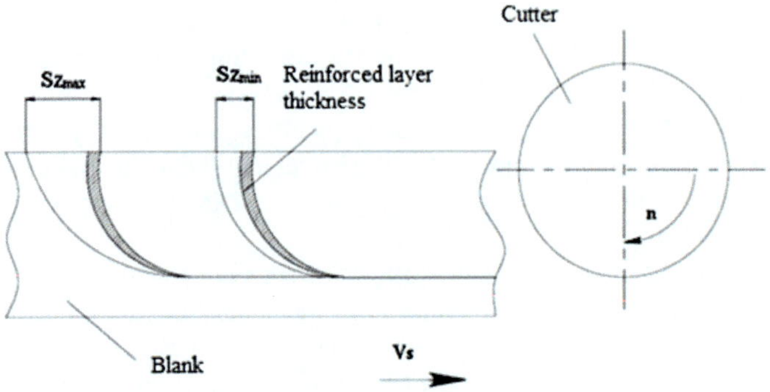

Fig. 5 Effect of shear rate Sz on components of shear forces at different thicknesses of the shear layer

is reduced. Adhesion of the processed material to the working surface of the milling cutter was evaluated by SteREO Discovery V12 stereomicroscopy. The radius r of the cutting edge rounding (undercut) was measured in MicroCAD lite Fig. 6. Change in the radius of penetration of the cutting blade ρ after 5 min of operation of the gasket is shown in Fig. 8 (Fig. 7).

It is known that when milling workpieces, wear of cutting tools occurs faster than when milling, since the cutting tool works under difficult conditions, and the teeth repeatedly hit the workpiece and enter and exit. The geometric parameters of a special cutting tool are characterized by the following geometric shapes: macrogeometry (drunk, compound, ovalrogomethy and microgegaometry (vegetable and pond). According to the general laws of erosion, sliding erosion of the cutting tool in the initial period of work is called initial erosion, and fast erosion occurs. In the initial period of wear, the cutting edge of the tool is machined, individual irregularities break up and migrate, and various lines, sharpening marks of the cutting edges are cleaned off and become smooth. This period of treated surfaces gadir is all slowly diminished [17–22].

Before applying a wear-resistant coating, the formation of rounding (lunch) in a disk electron-beam alloyed cutter is slower, and the cutting tool withstands high cutting forces, which slows down the onset of the rapid wear process. Cracks do not form on the cutting tool. If its partial rot is observed, it will quickly char and sharpen.

4 Conclusions

The results of the experiment indicate the possibility of obtaining modified surface layer, ion-nitriding, ion-alloying layers on the surface of a tool made of high-speed steel. These layers were obtained by initiating exothermic chemical reactions between the substrate and a 4 μm thin film deposited on it. In the reaction products,

Table 3 Wear of mills during machining

No	Name of the tool	Cutting modes			Tool durability T, min
		N, rpm	t, mm	S, mm/min	
1	Cut-off disk mill No. 1 ((TiCr)N-(TiAl)N-(CrAlSi)N + ion-nitriding covering)	350	4.0	150	1. Before machining the part, the outer diameter of the cutter was Ø99.5 mm
					2. Depth of cut 4 mm for a length of 200 mm
					3. After processing, deterioration on the outer diameter was 0.3 mm
2	Cut-off disk mill No. 2 (diamond-like covering (DLC))	350	4.0	150	1. Before processing, the cutting tool and the outer diameter of the cutter were Ø99.3 mm
					2. Depth of cut 4 mm for a length of 200 mm
					3. After processing, deterioration on the outer diameter was 0.25 mm
3	Cut-off disk mill No. 3 (((TiAl)N + ion-nitriding + ion-alloying (NbHf) covering)	350	4.0	150	1. Before processing, the cutting tool and the outer diameter of the cutter were Ø100 mm
					2. Depth of cut 4 mm for a length of 200 mm
					3. After processing, deterioration on the outer diameter was 0.2 mm
4	Cut-off disk mill No. 4 ((TiAl)N covering)	350	4.0	150	1. Before processing, the cutting tool and the outer diameter of the cutter were Ø99.5 mm
					2. Depth of cut 4 mm for a length of 200 mm
					3. After processing, deterioration on the outer diameter was 0.5 mm
5	Cut-off disk mill No. 5 ((TiAl)N + ion-nitriding covering)	350	4.0	150	1. Before processing, the cutting tool and the outer diameter of the cutter were Ø99.6 mm
					2. Depth of cut 4 mm for a length of 200 mm

(continued)

Table 3 (continued)

No	Name of the tool	Cutting modes			Tool durability T, min
		N, rpm	t, mm	S, mm/min	
					3. After processing, deterioration on the outer diameter was 0.3 mm
6	Cut-off disk mill No. 6 ((TiCr)N-(TiAl)N-(CrAlSi)N (nATCRo^3) covering)	350	4.0	150	1. Before processing, the cutting tool and the outer diameter of the cutter were Ø100.7 mm
					2. Depth of cut 4 mm for a length of 200 mm
					3. After processing, deterioration on the outer diameter was 0.3 mm
7	Cut-off disk mill No. 7 (without covering)	350	4.0	150	1. Before processing, the cutting tool and the outer diameter of the cutter were Ø99.5 mm
					2. Depth of cut 4 mm for a length of 200 mm
					3. After processing, deterioration on the outer diameter was 0.6 mm

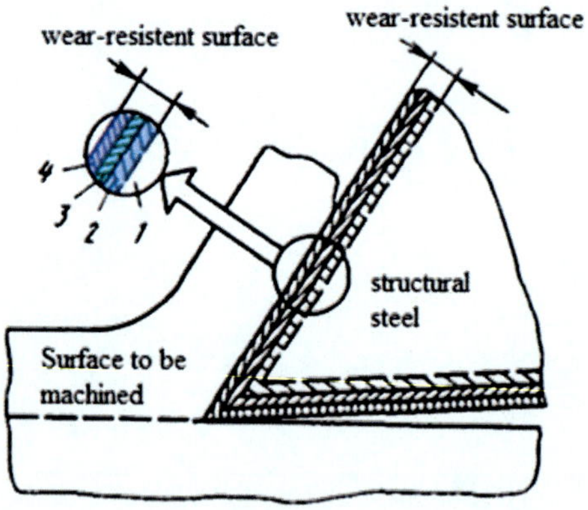

Fig. 6 Complicated machined surface. 1-tool steel, 2-adhesive heat-resistant nitrided layer, 3-alloyed base, 4-covering

the formation of components of a new phase has been determined. The implementation of such complex processing processes, ion nitriding, ion alloying of the base casting of cutting tools reduces its plastic deformation.

There are a large number of methods for obtaining coverings on the working surfaces of cutting tools. Physical precipitation, the covering of this machining on the

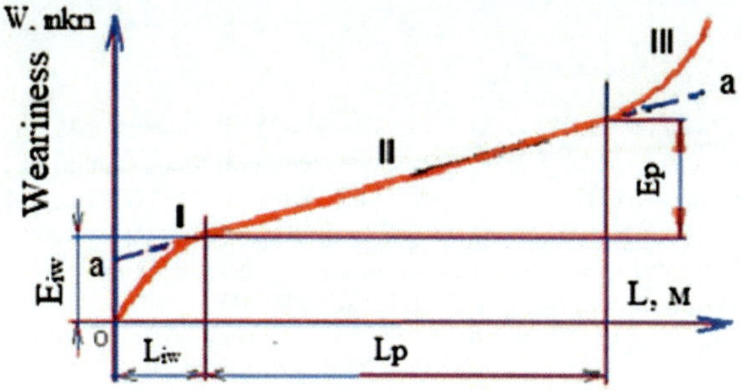

Fig. 7 Graph of cutting tool wear versus cutting path. L_{iw}-initial wear path, E_{iw}-initial wear, Lp-proportional length, Ep-proportional weariness

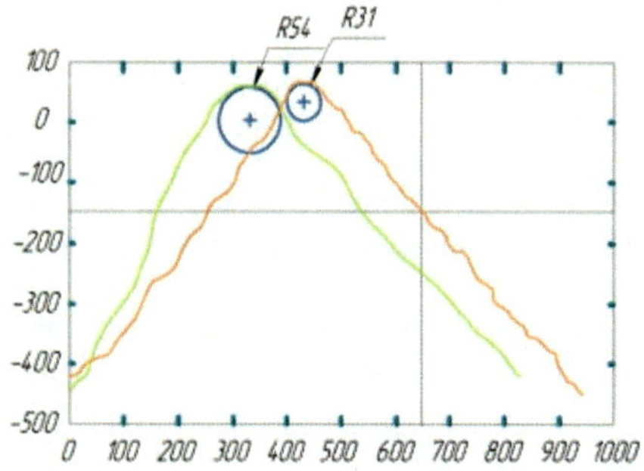

(a) - (TiAl) N coated cutting tool

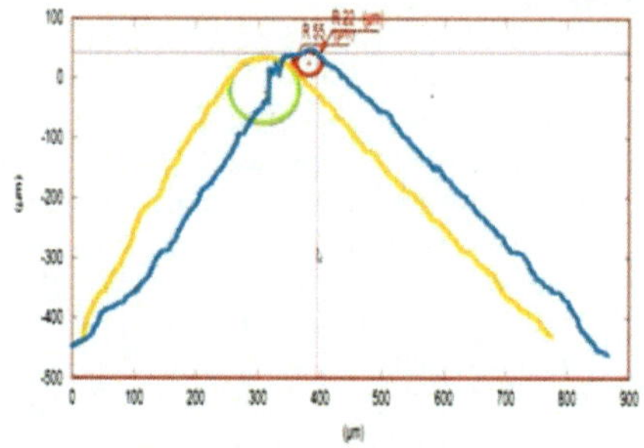

(b) - is a complex processed tool

Fig. 8 Change in the radius of penetration of the cutting blade ρ after 5 min of operation of the gasket. (3D scanner GFM)

surface of a tool material, which significantly differs in its crystal-chemical, physical–mechanical and thermal-physical properties from the corresponding properties of the tool material (base), significantly improves the properties of the latter.

Ion-nitriding and microalloying may also be recommended. In particular, such treatment, which is recommended to be carried out before applying a wear-resistant coating, makes it possible to influence the wear processes of not only high-speed but also high-speed tools, which makes it possible to increase the wear resistance of the tool by 3–4 times.

This surface slows down the growth process in the cutting zone during processing with complex cutting tools and helps to remove the generated heat without absorbing it. The relevance of the study lies in the fact that one of the factors affecting the quality of the manufactured product is the inaccuracy of cutting tools, which in many cases, is transferred directly to the workpiece, causing systematic errors in the shape and size of the surfaces of the workpieces. Based on the main task of the research, the plastic deformation of the coating on the surface of cutting tools working at high loads was reduced by chemical-thermal treatment and coating.

References

1. Fedorov SV, Aleshin SV, Swe MH, Abdirova RD, Kapitanov AV, Egorov SB (2017) Comprehensive surface treatment of high-speed steel tool. Mech Indus 18:711
2. Sharipov J, Barakayev F, Fozilov S, Karimova Z, Zaripov M (2022) Increasing the resistance of the cutting tool during heat treatment and coating. AIP Conf Proc 2432:050042
3. Fedorov S, Sharipov J, Abrorov A (2021) Increasing the surface stability of the cutting tool through complex machining. J Phys Conf Ser 1889:022079
4. Vereschaka AA, Grigoriev SN (2017) Study of cracking mechanisms in multi-layered composite nano-structured coatings. Wear 378–379:43–57. https://doi.org/10.1016/j.wear.2017.01.101/
5. Vereschaka A, Volosova M, Chigarev A, Sitnikov N, Ashmarin A, Sotova C, Bublikov J, Lytkin D (200) Influence of the thickness of a nanolayer composite coating on values of residual stress and the nature of coating wear. Coatings 10:63.https://doi.org/10.3390/coatings10010063
6. Xu YX, Riedl H, Holec D, Chen L, Du Y, Mayrhofer PH (2017) Thermal stability and oxidation resistance of sputtered Tisingle bondAlsingle bondCrsingle bondN hard coatings. Surf Coat Technol 324:47–57
7. Abrorov A, Kuvoncheva M, Rajabov O, Mukhammadov M, Jumaev S (2020) Method of thermal treatment of saw disk teeth of fiber-processing machines by laser quenching. IOP Conf Ser Mater Sci Eng 862:032034
8. Li WZ, Liu HW, Evaristo M, Polcar T, Cavaleiro (2013) A Influence of Al content on the mechanical properties and thermal stability in protective and oxidation atmospheres of Zr-Cr-Al-N coatings. Surf Coat Technol 236:238–246
9. Holec D, Rachbauer R, Chen L, Wang L, Luef D, Mayrhofer PH (2011) Mayrhofer phase stability and alloy-related trends in Ti–Al–N, Zr–Al–N and Hf–Al–N systems from first principles. Surf Coat Technol 206:1698–1704
10. Tabakov VP, Sagitov DI (2015) The performance of the cutting tool with Wear-resistant coatings in conditions of constrained cutting. UlGTU, Ulyanovsk, p. 179
11. Tabakov VP (2008) Formation of wear-resistant ion-plasma coatings of cutting tools. M, Mashinostroenie, p. 311
12. Vereshchaka AA (2010) Improving the cutting properties of carbide tools by rational choice of the composition, structure and properties of nanosized wear-resistant complexes: Dis. for the competition uch. step. cand. tech. Sciences. GOU VPO MSTU "Stankin", M, p. 254
13. www.mdpi.com
14. Rakhmonov K, Fayziev S, Rakhimov K, Kazakova D (2021) Relative speed and temperature effect investigation of the of the drying agent on the moisture content of cotton. E3S Web Conf 264:04008
15. Díaz-Guillén JC, Naeem M, Hdz-García HM, Acevedo-Davila JL, Díaz-Guillén MR, Khan MA, Iqbal J, Mtz-Enriquez AI (2020) Duplex plasma treatment of AISI D2 tool steel by combining plasma nitriding (with and without white layer) and post-oxidation. Surf Coat Technol 385:125420
16. Krysina OV, Prokopenko NA*, Ivanov YF, Tolkachev OS, Shugurov VV, Petrikova EA (2020) Multi-layered gradient (Zr,Nb)N coatings deposited by the vacuum-arc method. Surf Coat Technol 393:125759
17. Yamamoto K, Sato T, Takahara K, Hanaguri K (2003) Properties of (Ti,Cr,Al)N coatings with high Al content deposited by new plasma enhanced arc-cathode. Surf Coat Technol 174–175:620–626
18. Huang F, Wei G, Barnard JA, Weaver ML (2001) Microstructure and stress development in magnetron sputtered TiAlCr(N)films. Surf Coat Technol 146–147:391–397
19. Donohue LA, Smith IJ, Munz W-D, Petrov I, Greene JE (1997)"Surf. Coat Technol 94–95:226
20. Tam PL, Zhou ZF, Shum PW, Li KY (2008) Structural, mechanical, and tribological studies of Cr-TiAl-N coating with different chemical compositions. Thin Solid Films 516(16):5725–5731

21. Xu YX, Riedl H, Holec D, Chen L, Du Y, Mayrhofer PH (2017) Thermal stability and oxidation resistance of sputtered Tisingle bondAlsingle bondCrsingle bondN hard coatings. Surf Coat Technol 324:48–56
22. Vereschaka A, Volosova M, Chigarev A, Sitnikov N, Ashmarin A, Sotova C, Bublikov J, Lytkin D (2020) Influence of the thickness of a nanolayer composite coating on values of residual stress and the nature of coating wear. Coatings 10:63.https://doi.org/10.3390/coatings10010063

Chapter 7
Nano-Severe Plastic Deformation Process on Aluminum Alloy and Composites: An Overview

K. G. Sagar

Abstract Material when subjected to severe plastic deformation at the nanoscale level in a process known as nano-severe plastic deformation (NSPD), which produces a nanostructured material. Materials with ultrafine grain sizes, high strength, and enhanced mechanical properties can now be created employing NSPD processes. To create nanocrystalline and ultrafine-grained materials, the process involves subjecting the material to high plastic strains to cause deformation at the sub-micron level. Mechanical and thermo-mechanical processes are the two broad groups into which NSPD approaches can be categorised. High-pressure torsion, equal channel angular pressing, repeating corrugation, and straightening are examples of mechanical processes, while accumulative roll bonding and cyclic extrusion compression are examples of thermo-mechanical processes. A variety of nanostructured materials, including metals, alloys, and composites, have been created using NSPD. These materials may find use in a variety of industries, including those that call for high-strength and lightweight materials, like aerospace, automotive, and biomedical engineering. High plastic strains, often between 100 and 1000%, are applied to the material during NSPD, which causes the production of dislocations, grain boundary movement, and deformation twinning. As a result, mechanical qualities such as high strength, ductility, and fatigue resistance are enhanced. In this work, a review of the effect of NSPD on aluminium and aluminium metal matrix composites have been dealt.

Keywords Severe plastic deformation · Nano severe plastic deformation · Grain refinement · Mechanical and tribological properties

K. G. Sagar (✉)
Department of Mechanical Engineering, Cambridge Institute of Technology, Bengaluru, Karnataka, India
e-mail: sagar.mech@cambridge.edu.in

V. Bindhu et al. (eds.), *Proceedings of Sixth International Conference on Inventive Material Science Applications*, Advances in Sustainability Science and Technology,
https://doi.org/10.1007/978-981-99-4189-6_7

1 Introduction

A class of materials known as aluminum alloys includes aluminium as the main component as well as silicon, copper, magnesium, zinc, and other alloying elements. Because of their low density, high strength-to-weight ratio, excellent corrosion resistance, and superior formability, these alloys are widely employed in a wide range of applications. Cast alloys and wrought alloys are the two basic categories of aluminium alloys. Cast alloys are created by pouring the molten metal into a mold, whereas wrought alloys are created by rolling, extruding, forging, or other forming techniques [1].

These pure aluminium alloys, known as the 1000 family, are frequently employed for their exceptional formability and corrosion resistance. They are frequently employed in the production of cooking utensils, electrical conductors, and packaging materials. Copper serves as the main alloying element in the 2000 series alloys, which are renowned for their high strength and superior fatigue resistance [2–8]. They are frequently utilised for structural components in the aircraft sector. Manganese is the main alloying element of the 3000 series alloys, which are renowned for their good corrosion resistance and average strength. They are frequently employed in the production of automobile components and heat exchangers. Magnesium is the main alloying element of the 5000 series alloys, which are renowned for their high strength and superior corrosion resistance [2–8]. They are frequently employed in the production of maritime applications and automotive parts. Magnesium and silicon serve as the principal alloying components in the 6000 series alloys, which are renowned for their good strength, corrosion resistance, and formability. They are frequently employed in the production of automotive and architectural elements. Zinc serves as the main alloying element in the 7000 series alloys, which are renowned for their high strength and superior fatigue resistance. They are frequently utilised for structural components in the aircraft sector. A wide range of industries, including aerospace, automotive, construction, and consumer goods, employ aluminium alloys extensively. Current studies are concentrated on creating novel alloys with enhanced characteristics, streamlining production methods, and investigating new uses for these materials [2–8].

1.1 Aluminum Metal Matrix Composites (AMMCs)

Aluminum serves as the matrix material for aluminium metal matrix composites (AMMCs), a class of materials that also include one or more reinforcement materials such as ceramic particles, fibres, or whiskers. AMMCs combine high stiffness, strength, and wear resistance of the reinforcement elements with the beneficial features of aluminium, such as high strength-to-weight ratio and good corrosion resistance. Ceramic particles, fibres, and whiskers are the three main categories for the reinforcement materials utilised in AMMCs [9–13]. The most popular type of reinforcement material in AMMCs is ceramic particles because of their low price, convenient availability, and superior mechanical qualities. On the other hand, fibres and whiskers are more expensive but have better mechanical qualities. The preparation of the reinforcement material, mixing of the reinforcement material with the aluminium matrix material, and consolidation of the mixture are all phases in the manufacture of AMMCs. Many methods, including powder metallurgy, casting, and infiltration, can be used to consolidate. AMMCs have stronger stiffness, strength, and wear resistance than typical aluminium alloys, among other benefits. They are frequently utilised in areas including aerospace, automotive, and military, where strong mechanical characteristics and wear resistance are necessary [9–13]. AMMCs are frequently used in the manufacture of brake discs, engine parts, and structural elements for airplanes [9–13]. The goal of the ongoing study is to enhance the properties of AMMCs while also studying new types of reinforcing materials, developing innovative processing methods, and expanding the scope of their industrial uses. So, in this work, the author is discussing the nano-SPD of aluminum and aluminium metal matrix composites.

1.2 Nano-Severe Plastic Deformation (NSPD)

A material is subjected to severe plastic deformation in the nanoscale range in a process known as nano-severe plastic deformation (nano-SPD). By reducing the grain size to the nanoscale range, this method aims to enhance the material's mechanical and physical characteristics, such as strength, ductility, and wear resistance. High-pressure torsion (HPT), equal channel angular pressing (ECAP), and friction stir processing are just a few of the methods employed for nano-SPD (FSP) [14]. To achieve the appropriate nanoscale grain structure, each of these methods deforms the material under intense pressure and shear strain. Improved strength and ductility, greater wear resistance, and higher thermal stability are all advantages of nano-SPD [14, 15]. These enhancements result from the finer grain size, which increases the density of grain boundaries and dislocations and lessens the likelihood of material faults. Aerospace, automotive, and biomedical are just a few of the numerous areas where nano-SPD could find useful. Nano-SPD can be utilised, for instance, to make stronger and more resilient structural elements for cars and airplanes, as well as to increase the wear resistance of medical implants. The mechanical and tribological characteristics of aluminium alloys have been discovered to be significantly impacted by nanoscale severe plastic deformation (nano-SPD). Strength, ductility, wear resistance, and fatigue life are all enhanced by the nanoscale grain refinement achieved with nano-SPD. One study looked at how ECAP affected an aluminium alloy's mechanical characteristics. The scientists discovered that after ECAP, the material's strength and hardness greatly enhanced, while its ductility remained mostly unchanged. The increase in strength was attributable to the nanoscale grain size refinement, which raised the density of dislocations by increasing the number of grain boundaries. Another study investigated how HPT affected the tribological characteristics of an aluminium alloy. The creation of a homogenous, refined microstructure with a high density of dislocations and grain boundaries after HPT was found to greatly boost the material's wear resistance. The development of a stable tribolayer on the material's surface was also noted by the researchers as a decrease in the friction coefficient and wear rate [14, 15]. Nano-SPD can increase the thermal stability of aluminium alloys in addition to enhancing their mechanical and tribological characteristics. For instance, one study discovered that ECAP increased the thermal stability of the grain boundaries, which in turn improved the high-temperature stability of an aluminium alloy. Fatigue life improvement: Nano-SPD has been found to significantly improve the fatigue life of aluminium alloys [14, 15]. One study found that after ECAP, the fatigue strength of an aluminium alloy increased by up to 60%. The improvement in fatigue life was attributed to the refinement of the grain size, which led to a greater number of dislocations and a higher density of grain boundaries, both of which contribute to improved fatigue resistance. Corrosion resistance: Nano-SPD has also been found to improve the corrosion resistance of aluminium alloys. One study found that after HPT, the corrosion resistance of an aluminium alloy was significantly improved due to the formation of a homogenous and refined microstructure with a high density of grain boundaries and dislocations. Microstructure refinement: One of the primary effects of nano-SPD on aluminium alloys is the refinement of the microstructure to the nanometre scale. This refinement leads to a more uniform distribution of the alloying elements and a higher density of grain boundaries and dislocations, which in turn leads to improvements in mechanical and tribological properties. Strain hardening: Nano-SPD induces a high level of strain hardening in aluminium alloys, which further contributes to improvements in mechanical and tribological properties. Strain hardening is the process by which the material becomes stronger and more resistant to deformation because of plastic deformation. Grain boundary strengthening: Using nano-SPD to refine the grain size to the nanoscale, the quantity of grain boundaries in the material is increased. The strengthening action that follows, known as grain boundary strengthening, helps to improve the mechanical properties of aluminium alloys. Aluminum alloys' texture, or the preferred orientation of the grains within the substance, can be changed through nano-SPD. The material's anisotropy, or directional dependency of characteristics, can be impacted by texture. A more random texture produced by nano-SPD may result in improved isotropic characteristics. Recrystallisation suppression: The material may recrystallise during typical plastic deformation, such as rolling or extrusion, which can result in coarsening of the grain size and a decrease in the strengthening effect. Yet, it has been discovered that nano-SPD inhibits recrystallisation, preserving the fine grain size and the strengthening effect. Fabrication of composites: Aluminum matrix composites with enhanced mechanical properties can also be created using nano-SPD. For instance, one study discovered that adding ceramic nanoparticles to an aluminium matrix via nano-SPD resulted in appreciable gains in strength and wear resistance. Surface modification: Nano-SPD can be used to modify the surface properties of aluminium alloys. For example, one study found that after HPT, the surface hardness of an aluminium alloy increased significantly, which was attributed to the refinement of the grain size and the increase in the density of dislocations. Improved wear resistance: Nano-SPD has been found to significantly improve the wear resistance of aluminium alloys. One study found that after ECAP, the wear resistance of an aluminium alloy increased by up to 300%, which was attributed to the refinement of the grain size and the increase in the density of dislocations and grain boundaries. Improved mechanical properties at high temperatures: Nano-SPD can also improve the mechanical properties of aluminium alloys at high temperatures. One study found that after HPT, the yield strength and tensile strength of an aluminium alloy increased significantly at high temperatures, which was attributed to the refinement of the microstructure and the suppression of recrystallisation. Improved formability: Nano-SPD can also improve the formability of aluminium alloys, which is their ability to be shaped into complex geometries without cracking or rupturing. One study found that after ECAP, the formability of an aluminium alloy increased significantly due to the refinement of the microstructure and the increase in the density of dislocations. Nano-SPD can also enhance the joining characteristics of aluminium alloys. For instance, one study discovered that the improvement in the microstructure and the rise in dislocation density during HPT considerably boosted the joint strength of an aluminium alloy. Improved electrical and thermal conductivity: Aluminum alloys continue to have strong electrical and thermal conductivities despite the grain refinement caused by nano-SPD, making them a desirable material for a variety of applications [15].

Aluminum metal matrix composites (AMMCs) can use a variety of reinforcing types, such as: Ceramic particles: Due to their high strength and stiffness, ceramic particles including silicon carbide (SiC), alumina (Al2O3), and titanium diboride (TiB2) are frequently employed as reinforcement in AMMCs. Metallic particles: AMMCs can also be reinforced with metallic particles including copper (Cu), nickel (Ni), and aluminium oxide (Al2O3). Carbon fibres: Carbon fibres are an appealing reinforcement for AMMCs because of their lightweight and excellent tensile strength [16–23]. Graphene: Because of its distinct mechanical, thermal, and electrical characteristics, graphene is a desirable reinforcement material for AMMCs. Nanoparticles: AMMCs can be reinforced with nanoparticles including carbon nanotubes (CNTs), graphene oxide (GO), and titanium oxide (TiO2). The mechanical and tribological characteristics of AMMCs can be greatly enhanced by these particles, which have a large surface area to volume ratio. Fibers: AMMCs can also be reinforced with fibres, including glass, aramid, and carbon fibres. These fibres can increase the stiffness and strength of AMMCs, which makes them perfect for applications requiring high strength. Whiskers: Long, thin fibres, like silicon carbide whiskers, can be employed as reinforcement in AMMCs. The mechanical qualities of AMMCs can be considerably enhanced by whiskers, which have high strength and stiffness [16–23].

High-pressure torsion (HPT) and equal channel angular pressing are the two most popular nano-SPD processes that can be utilised to process AMMCs (ECAP). In HPT, a sample is subjected to high pressure and torsion deformation, which causes a severe plastic deformation that can improve the material's microstructure. ECAP includes running a sample through a die that has two channels that cross at a precise angle.

This causes a significant plastic deformation that can also improve the material's microstructure. It has been discovered that processing different kinds of AMMCs with HPT and ECAP greatly enhances their mechanical and tribological properties. Several research have investigated the impact of nano-SPD on aluminium alloys reinforced with different types of reinforcements. These are a few instances.

Carbon nanotube-reinforced aluminium matrix composites: Scientists have discovered that introducing carbon nanotubes (CNTs) and then subjecting the composite to nano-SPD can greatly improve its mechanical characteristics.

Graphene oxide has also been employed as a reinforcement in aluminium matrix composites, and its utilisation is described below. Researchers have discovered that applying nano-SPD to Al6061/GO composites can greatly enhance their mechanical characteristics. It is an emerging method.

1.3 Advantages and Limitations of Nano-SPD

Increased strength, ductility, and hardness are just a few examples of the mechanical attributes that nano-SPD can help materials possess. The creation of ultrafine or even nanocrystalline structures in materials because of nano-SPD might enhance their microstructure and, consequently their mechanical characteristics. Versatility: Metals, alloys, and composites are just a few of the many materials to which nano-SPD can be used. Cost-effectiveness: Nano-SPD does not require expensive equipment or supplies, making it a comparatively inexpensive approach to enhancing the mechanical characteristics of materials.

1.3.1 Limitations of Nano-SPD

In summary, the SPD process has some limitations that can affect its scalability, cost-effectiveness, and accessibility. However, with continued research and development, some of these limitations can be addressed, leading to the development of new and improved SPD techniques for manufacturing materials with tailored properties.

1.4 Nano-SPD

The use of nano-SPD on a wider variety of materials, such as metals, ceramics, polymers, and composites, may be investigated by researchers. Customised microstructures: To maximise the characteristics of materials for purposes, researchers may concentrate on creating novel techniques for controlling and modifying the microstructure of materials utilising nano-SPD. Multi-scale processing: To produce materials with multi-scale microstructures, which might result in superior mechanical and functional properties, researchers may investigate the use of nano-SPD in combination with other processing methods. Large-scale manufacturing: To enable the commercial manufacture of high-performance materials, researchers may concentrate on creating large-scale manufacturing techniques for nano-SPD. Improved characterisation: To better understand the underlying principles and optimise the procedure for particular applications, researchers may develop novel approaches for assessing the microstructure and characteristics of materials made via nano-SPD. Simulation and modelling: To simulate and forecast the microstructural evolution and mechanical properties of materials produced with nano-SPD, researchers may create computational models. Advanced applications: Researchers might concentrate on creating cutting-edge uses for materials created via nano-SPD, like energy storage, aeronautical, and biological ones. Integration with additive manufacturing: Nano-SPD and additive manufacturing can be used together to produce materials with intricate geometries and microstructures, opening new design options for high-performance parts. Materials with various functions: To support the creation of cutting-edge technology, researchers may investigate the use of nano-SPD to produce materials with traits like high strength, thermal conductivity, and electrical conductivity. Environmental factors: Future studies may concentrate on enhancing the environmental sustainability of nano-SPD, for example, by lowering the amount of energy consumed and waste produced during the procedure. Commercialisation: As nano-SPD develops, there may be a greater interest in making the technology available for industrial uses, which could result in the creation of new markets and products. Integration with Industry 4.0: The combination of nano-SPD with Industry 4.0 technologies, including as artificial intelligence, machine learning, and the internet of things, may make it possible to optimise process parameters and boost the final product's quality and reproducibility. High-pressure torsion and accumulative roll bonding are two SPD techniques that could be combined with nano-SPD to develop hybrid processing methods that give distinctive microstructures and enhanced characteristics.

2 Conclusions

Nano-SPD is an effective method for enhancing the mechanical and tribological characteristics of aluminium alloys and composites made of aluminium metal matrix. The strength, ductility, hardness, and wear resistance of these materials can be considerably increased by nano-SPD through grain refinement and the creation of sub microscopic structures. These qualities can be improved even more by using other reinforcements, like nanoparticles and fibres. Although there are significant drawbacks and difficulties with nano-SPD, such as the requirement for specialised tools and the possibility of processing-induced errors, continuing research is resolving these problems and enhancing the technique's capabilities. In general, nano-SPD holds the promise of revolutionising the creation of innovative materials for a variety of industries, including aerospace, automotive, and biomedical engineering. We may anticipate many exciting new advancements in the field as researchers work to fully utilise this technology.

References

1. Types of Aluminum and Aluminum Alloy You Should Know | MachineMfg (2022) August 31, 2022. Www.machinemfg.com. https://www.machinemfg.com/series-of-aluminum-and-aluminum-alloy/
2. Aluminum and Aluminum Alloys nd. https://doi.org/10.1361/autb2001p351.
3. Dhanasekaran S, Sunilraj S, Ramya G, Ravishankar S (2015) SiC and Al2O3 reinforced aluminum metal matrix composites for heavy vehicle clutch applications. Trans Indian Inst Met 69(3):699–703. https://doi.org/10.1007/s12666-015-0542-8
4. Edalati K, Bachmaier A, Beloshenko VA, Beygelzimer Y, Blank VD, Botta WJ, Bryła K et al (2022) Nanomaterials by severe plastic deformation: review of historical developments and recent advances. Mater Res Lett 10(4):163–256. https://doi.org/10.1080/21663831.2022.2029779
5. Moona G, Walia R, Rastogi V, Sharma R (2018) Aluminium metal matrix composites: a retrospective investigation. Indian J Pure Appl Phys. https://www.semanticscholar.org/paper/Aluminium-metal-matrix-composites%3A-A-retrospective-Moona-Walia/52820d12159b65f427f1f96f1cceec6b830b8230
6. Garg P, Jamwal A, Kumar D, Sadasivuni KK, Hussain CM, Gupta P (2019) Advance research progresses in aluminium matrix composites: manufacturing & applications. J Market Res 8(5):4924–4939. https://doi.org/10.1016/j.jmrt.2019.06.028
7. Georgantzia E, Gkantou M, Kamaris GS (2021) Aluminium alloys as structural material: a review of research. Eng Struct 227(January):111372. https://doi.org/10.1016/j.engstruct.2020.111372
8. Joseph OO, Olubambi PA, Joseph OO, Edun BM, Okeniyi JO, Abioye OP (2021) Effects of alloying on aluminium-silicon alloys–a review. IOP Conf Ser Mater Sci Eng 1107(1):012116. https://doi.org/10.1088/1757-899x/1107/1/012116
9. Kotteda TK, Kumar M, Kumar P, Chekuri RBR (2022) Metal matrix nanocomposites: future scope in the fabrication and machining techniques. Int J Adv Manuf Technol. https://doi.org/10.1007/s00170-022-09847-0
10. Kumar H, Devade K, Singh DP, Giri JM, Kumar M, Arun V (2023) Severe plastic deformation: a state of art. Mater Today Proceed. https://doi.org/10.1016/j.matpr.2023.02.194
11. Mavhungu ST, Akinlabi ET, Onitiri MA, Varachia FM (2017) Aluminum matrix composites for industrial use: advances and trends. Proced Manuf 7:178–182. https://doi.org/10.1016/j.promfg.2016.12.045
12. Metal Matrix Composite-an Overview | ScienceDirect Topics. nd. Www.sciencedirect.com. Accessed April 14, 2023. https://www.sciencedirect.com/topics/materials-science/metal-matrix-composite#:~:text=Aluminum%E2%80%93matrix%20composites%20are%20most%20commonly%20studied%20MMC%20as
13. Pasha MBA, Kaleemulla M (2018) Processing and characterization of aluminum metal matrix composites: an overview. Rev Adv Mater Sci 56(1):79–90. https://doi.org/10.1515/rams-2018-0039
14. Rajaganapathy C, Vasudevan D, Selvakumar N (2020) Investigation on tribological and mechanical behaviour of AA6082—graphene based composites with Ti particles. Mater Res Express 7(7):076514. https://doi.org/10.1088/2053-1591/aba508
15. Saini P, Singh PK (2022) Fabrication and Characterization of SiC-reinforced Al-4032 metal matrix composites. Eng Res Express 4(1):015004. https://doi.org/10.1088/2631-8695/ac4831

16. Santos MC, Machado AR, Sales WF, Barrozo MAS, Ezugwu EO (2016) Machining of aluminum alloys: a review. Int J Adv Manuf Technol 86(9–12):3067–3080. https://doi.org/10.1007/s00170-016-8431-9
17. Singh L, Kumar S, Shivam SR, Badhani P (2021) Aluminium metal matrix composites: manufacturing and applications. IOP Conf Ser Mater Sci Eng 1149(1):012025. https://doi.org/10.1088/1757-899x/1149/1/012025
18. Subramanian C (2010) Wear properties of aluminium-based alloys. Surf Eng Light Alloys 40–57.https://doi.org/10.1533/9781845699451.1.40
19. Sundar G, Rajesh Jesudoss Hynes N (2019) Reinforcement in aluminium metal matrix composites. Advances in basic science (ICABS 2019). https://doi.org/10.1063/1.5122398
20. Types of Aluminum: Types, Uses, Features and Benefits nd. Accessed April 14, 2023. Www.iqsdirectory.com. https://www.iqsdirectory.com/articles/aluminum-metal/types-of-aluminum.html#:~:text=The%201000%20series%20is%20non-heat%20treatable%20and%20contains
21. Vijaya Ramnath B, Parswajinan C, Dharmaseelan R, Thileepan K, Nithin Krishna K (2021) A review on aluminium metal matrix composites. Mater Today Proceed 46:4341–43. https://doi.org/10.1016/j.matpr.2021.03.600
22. Yashpal S, Jawalkar CS, Verma AS, Suri NM (2017) Fabrication of aluminium metal matrix composites with particulate reinforcement: a review. Mater Today Proceed 4(2):2927–2936. https://doi.org/10.1016/j.matpr.2017.02.174
23. Zhang A, Li Y (2023) Thermal conductivity of aluminum alloys—a review. Materials 16(8):2972. https://doi.org/10.3390/ma16082972

Chapter 8
Review of Mechanized Garbage-Picking Machines: Challenges and Advancements in Sweeping and Handling of Trash

Nihal Wargantiwar, V. N. Bhaiswar, and V. W. Khond

Abstract The purpose of this paper is to study various mechanized garbage-picking machines with the view to overcome the challenges associated with roadside garbage picking, various sweeping machines, sweeping processes, handling garbage, and transportation of garbage. This study article summarizes the findings of numerous patents and studies done to improve various garbage-picking mechanisms. Garbage can be collected using a manual, electric motor-powered, or power-operated method. The first manually operated device is typically utilized to collect trash from minor roadways. To work, a second-powered mechanism needs gasoline or diesel. Electric battery power is used to operate the third electric power mechanism. Each mechanism has advantages and limitations. Comparative analyses of various mechanisms are carried out using prior research papers and patents. The majority of patients and researchers indicated enhanced performance and optimized mechanisms employing manual, electric, or other forms of power. These solutions have substantial maintenance costs as well, which raises the overall cost. The revised in this review article.

Keywords Sweeping machine · Garbage collection · Rotary brush · Trash · Trash container

N. Wargantiwar (✉) · V. N. Bhaiswar · V. W. Khond
Department of CAD/CAM, G H Raisoni College of Engineering, MTech, Nagpur, Maharashtra 440016, India
e-mail: wargantiwarnihal@gmail.com

V. N. Bhaiswar
e-mail: vinod.bhaiswar@raisoni.net

V. W. Khond
e-mail: vivek.khond@raisoni.net

V. Bindhu et al. (eds.), *Proceedings of Sixth International Conference on Inventive Material Science Applications*, Advances in Sustainability Science and Technology,
https://doi.org/10.1007/978-981-99-4189-6_8

1 Introduction

In India, garbage cleaning is one of the major problems which covers all major cities and villages, especially on roads, railway stations, airports, and bus stops [1]. Managing garbage in the nation is a huge challenge. Owing to increasing urbanization. The 62 million tons of rubbish are produced annually by over 377 million urban residents of 7,935 towns and localities [2]. That dust is not commonly cited as a cause of road accidents in police reports, making data linking the two scarce. However, studies suggest that if dust contributes to 10% of accidents in developing countries, it could cost up to $800 million annually and account for 0.02% of GDP, with 40% of dust emissions stemming from vehicle activities on the road [3]. Municipal corporations have been assigned by the government to handle critical services. India's trend is inconsistent. Few local government agencies decide to clean the roadways by themselves. Other municipal authorities desire to give sweeping private companies the entire contract for sweeping. Early in the morning, they hire employees to sweep the roadside trash, bottles, and leaf litter. However, there are numerous health and safety-related issues at the moment. Road sweeping is currently carried out by contractors in large urban areas. Indian road sweeping industry is still in its infancy, as seen by the effectiveness of the country's native roads. Road sweeping machines have given the impression that they are not appropriate for the relatively particular demands of Indian roads [4]. In India, cleaning roads with brooms and hand-held devices is time-consuming and unhealthy for operators due to dust inhalation and dehydration. The "Clean India Mission" has created a need for a more efficient and healthy trash-picking device [5]. Improper waste management due to rising population is a major issue. A planned system using ZigBee, GSM, and RFID technologies is designed for indoor floor waste collection in high-rise buildings [6].

Street sweeping is performed by 9231 regular employees of the Municipal Corporation of Greater Mumbai (MCGM), although this number must be significantly greater across all of India. Studies have found a link between occupational exposure to sweeping and the onset of chronic respiratory illnesses, skin conditions, eye irritation, asthma, TB, and hypertension among employees. Musculoskeletal injuries make up the majority of the other non-fatal injuries that were found [7]. Cleaners frequently labor hunched over and with their backs bowed. They carry out several repetitive activities every day [4].

Pain, soreness, or discomfort in any of the anatomical regions of the body, including the neck, shoulders, upper back, lower back, elbows, wrists, or hands, as well as the hips or thighs, knees, ankles, or feet, are referred to as musculoskeletal disorders (MSDs) [8]. Regular cleaning and sanitizing not only protect human health but also prevents pest infestations and improve the durability of surfaces [9].

- **Sweeping machines generally work on the following three principles**

The regenerative air sweeper: It is typically used for residential purposes on streets, parking lots, and back roads. It uses a closed loop of air without venting or fumes. It is the most adaptable advancement for getting ordinary street junk like overpowering earth, rocks, sand leaves, and whip. To clean, regenerative air sweepers use a closed-loop "cyclonic effect." They share a vacuum inlet with vacuum sweepers in the same location on one side of the sweeping head. However, in contrast to vacuum machines, regenerative air sweepers continuously regenerate their internal air supply [10].

Broom Sweeper Machines: Mechanical broom sweepers have a main drum-like broom that runs transversely from one side of the sweeper to the other so that the broom bristles contact the paved surface throughout the sweeper unit's width. This technology can be compared to cleaning with a broom and parker. An attached refuse collector or bucket is used to collect swept debris for subsequent delivery to the dump site [10].

Vacuum Sweepers: This may be likened to a household vacuum system. The blow that creates the necessary vacuum and suction pressure is powered by a motor. A relatively recent technology is the high-efficiency sweeper. It uses a variety of fugitive dust loss controllers and a wide range of diverse devices, including vacuums, regenerative air systems, and at least one well-known mechanical broom machine. These devices are made to limit fugitive dust losses in addition to removing a significant amount of collected material of all sizes, particularly small-micron particles of less than 60 μ [10].

- **Most commonly used method of waste disposal in India**

In India, the most commonly used method of waste disposal is still landfilling, although efforts are being made to shift towards more sustainable practices. According to a report by the Central Pollution Control Board, around 75% of the country's solid waste is still disposed of in landfills.

However, due to the increasing amount of waste generated and the limited availability of land for new landfills, the Indian government is encouraging the use of alternative waste management practices such as waste-to-energy, composting, and recycling.

Waste-to-energy involves converting waste into energy through various processes such as incineration or anaerobic digestion. Composting involves the decomposition of organic waste into nutrient-rich soil, which can be used as a fertilizer. Recycling involves processing waste materials into new products.

In recent years, various initiatives and policies have been implemented by the Indian government to promote sustainable waste management practices and reduce the amount of waste generated. For example, the Swachh Bharat Abhiyan (Clean India Mission) launched in 2014 aims to achieve the goal of a clean India by 2022 by promoting sustainable waste management practices and improving sanitation infrastructure.

- **Best disposal method for garbage that is eco-friendly**

The most eco-friendly method of waste disposal is to reduce the amount of waste generated in the first place, followed by reusing, recycling, and composting as much waste as possible. However, when waste cannot be reduced, reused, recycled, or composted, the following disposal methods are considered to be the most eco-friendly:

(1) Waste-to-energy (WTE): Waste-to-energy involves converting waste into energy through various processes such as incineration, gasification, and pyrolysis. These processes generate heat or electricity from the waste, which can then be used to power homes or businesses. While waste-to-energy has some environmental impacts, such as air emissions, it can reduce the amount of waste sent to landfills and offset the use of fossil fuels.
(2) Anaerobic digestion: Anaerobic digestion is a biological process that breaks down organic waste, such as food and yard waste, into biogas and nutrient-rich soil. The biogas can be used to generate electricity or heat, while the soil can be used as a fertilizer. Anaerobic digestion has a lower environmental impact than incineration and can help to reduce greenhouse gas emissions.
(3) Land application: Land application involves applying treated wastewater and biosolids to land for agricultural purposes. This method can provide a beneficial use for waste products while improving soil quality and reducing the need for chemical fertilizers. However, land application can also have negative environmental impacts, such as contaminating groundwater.

Overall, the most eco-friendly method of waste disposal will depend on the specific waste stream, local regulations, and available infrastructure. It is important to prioritize waste reduction, reuse, recycling, and composting, and to only use disposal methods that have been proven to be safe and effective.

Orders of waste handling methods:

The waste handling methods can be ordered in a hierarchy known as the "waste management hierarchy." This hierarchy lists the preferred methods for managing waste in order of priority, from the most to least environmentally friendly options. The waste management hierarchy typically includes the following methods, listed in the order of preference:

1. Source Reduction and Waste Prevention: This involves reducing the amount of waste generated in the first place by changing consumer behavior, redesigning products, and reducing packaging.
2. Reuse: This involves using materials and products multiple times before they are discarded, such as by repairing, refurbishing, or repurposing them.
3. Recycling and Composting: This involves processing waste materials to recover valuable resources, such as paper, plastic, and metals, or to turn organic waste into compost.
4. Recovery of Energy: This involves recovering energy from waste through processes such as incineration, gasification, and anaerobic digestion.
5. Treatment: This involves treating waste to reduce its volume and make it less hazardous before disposal.

6. Landfilling: This involves burying waste in a landfill, which is typically the least environmentally friendly waste management option.

The waste management hierarchy is a useful tool for guiding waste management decisions and promoting environmentally sustainable waste management practices. By following this hierarchy, we can prioritize the most sustainable waste management practices and reduce the amount of waste that ends up in landfills.

2 Literature Review

1. **Manually operated sweeper**

Michaels [11] invented the manual sweeping machine. This machine is an entirely manually operated and semi-automatic machine. This machine has a rotary brush, hopper, and large trash container. The sweeping mechanism is connected to a bicycle, and the operator needs to push the pedal of the bicycle to run this machine. As the operator moves the bicycle by pushing the sweeping pedal mechanism also moves in the same direction. As the machine moves, wheels are rotated, and a rotary brush is connected with the wheels from gears. As the wheels rotate, the rotary brush rotates anticlockwise direction and pushes entrapped garbage to the hopper, and the hopper conveys that garbage to a large collecting tank [11]. A schematic drawing of the machine shows in Fig. 1.

Limitations:

(1) This machine cannot pick the big type of garbage like water bottles, cans, cardboard, papers, etc.
(2) Hard to maneuver, handling this machine a very complicated task.
(3) Turning off this machine is very hard.

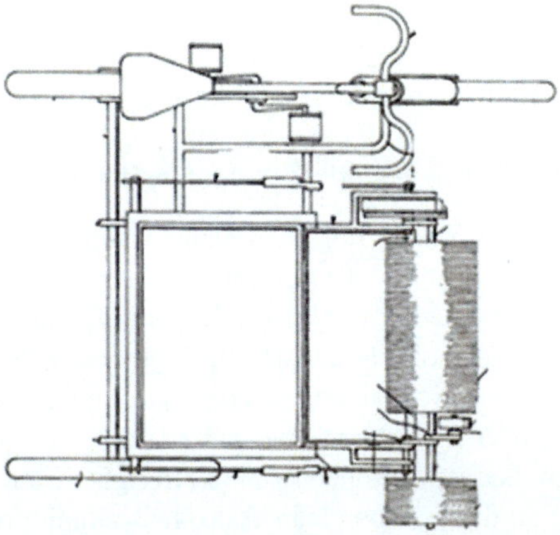

Fig. 1 Manual sweeper machine (Abraham Michaels, US3947912A) [11]

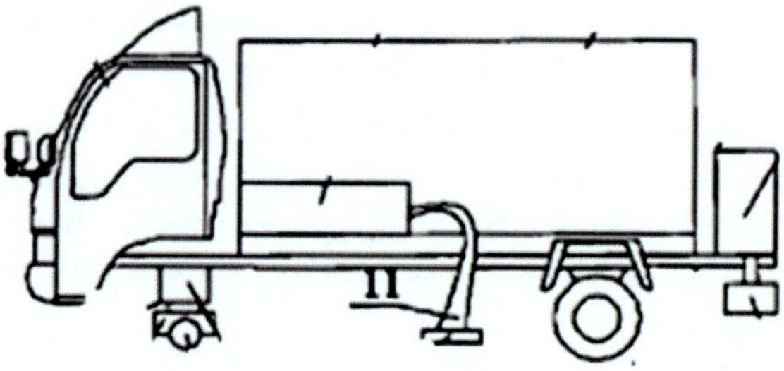

Fig. 2 Road refuse sweeping vehicle (项侠风(Xiang Xiafeng), CN203514238U) [12]

(4) Apart from garbage picking, many challenges to face the operator while moving in actual work conditions because maneuver is a very complicated task.

2. **Road refuse sweeping vehicle**

A 项侠风"(Xiang Xiafeng) invented the automobile that sweeps up road debris as seen in the utility model. The road refuses sweeping vehicle is made up of a chassis, a vehicle head, and a refuse tank, both of which are mounted to the chassis. On the front part of the chassis, there is a removal mechanism. In the middle of the chassis is a dust collection system. On the back of the chassis, there is a water spraying system. The removal device may successfully remove refuse that has stuck to the road. The dust collection device is capable of effectively sucking up road debris and dust, the debris can be sucked and collected in the collecting chamber. The water spraying mechanism efficiently maintains a clean surface on which dust has been vacuumed up. Additionally, the water spraying system may stop dust from scattering. The vehicle used for road waste sweeping keeps city streets tidy and appropriate for use and popularization [12]. A schematic drawing of the machine is shown in Fig. 2.

Limitations:

(1) Main disadvantage of this machine is that the operating cost of this machine is high.
(2) It is too big and it can cause traffic issues.
(3) Not compatible to work on narrow roads.
(4) Much power is required to move the entire sweeping mechanism.

3. **A kind of road that dust is dust removal integrated cleaning trolley**

A 郑佳 (Zheng Jia) [13] invented a kind of road that dust is a dust removal integrated cleaning trolley. The invention describes a type of roadway where the removal of dust is integrated with cleaning the trolley, including the locomotive and the placement of the disposal box inside the locomotive. Both a trash can and a water tank, moving pulleys are fitted in the locomotive's lower part. Pushing hands hold the locomotive's right end in place. The water tank, garbage collection box, and disposal box are gradually positioned from left to right inside the locomotive. The equipment collects trash that can be cleaned up. It is simple to use and increases the effectiveness of road cleaning, garbage is delivered to the device's garbage collection box using a rubbish conveyer belt. The leftover trash is gathered in a secondary garbage collection container. The disposal box gathers microscopic dust, water can be sprayed from a tank after cleaning with the water sprayer hole on the watering plate to the ground. The phenomenon that prevents dust from flying upward and is pollution-free. This machine improves the effectiveness of cleaning the road and is suitable for more smooth road surfaces. It replaces the manual sweeping of roads [13]. A schematic drawing of the machine is shown in Fig. 3.

Limitations:

(1) Main disadvantage of this machine is that the operating cost of this machine is high.
(2) Machine cost is high as it is constructed on locomotives.
(3) Much power is required to move the entire sweeping mechanism.

4. **Sweeping machine tractor attachment for sports grounds**

This mechanism was invented by Derek Burtt Hargreaves Adlington, near Macclesfield; Eric Stanforth, Macclesfield; Edward Hobbs, Poynton, all of England [14]. The frame of the sweeper attachment can be connected to a tractor's hydraulic lifting equipment. A cable is provided to tip the open-fronted collector within the frame, which is pivotally mounted. The collector swings back into position to receive litter due to gravity and is secured there by a releasable lock. The litter is swept into the collector by a rotating brush in front of it. The sweeper attachment is supported by relatively big wheels near the brush, and the attachment's back is supported by smaller casters. Adjustable ground wheels' height about the frame.

Sweepings collector tilt is ably mounted as a unit on the frame rearward of the sweeping mechanism and is movable relative to the frame between a normal or receiving position and a tilted position for discharging collected sweepings forwardly in the ground-engaging means; sweepings mechanism supported by the frame and provided with driving means; A sweeping apparatus for mounting on a tractor is also

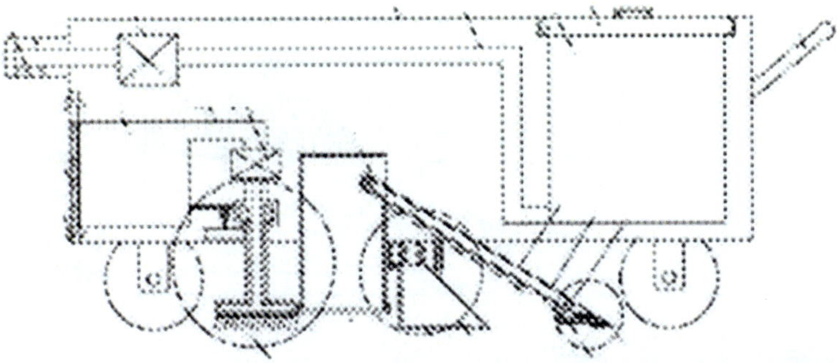

Fig. 3 A kind of road that dust is dust removal integrated cleaning trolley (郑佳 (Zheng Jia) CN108755529A) [13]

included in the invention. This apparatus has a frame with a front part and a rear part, a sweeping mechanism including a brush mounted on the front part and rotatable about an axis of rotation, ground wheels on the front part for providing support from the ground to the part adjacent to the brush, the coupling means for connecting the front part to the tractor, and ground wheels. Engagement mechanisms on the frames are in the back portion.

The rear half of the frame can move up and down in proportion to the front half as the sweeping machine is pushed across the uneven ground without appreciably affecting the height of the brush's axis of rotation above the ground [14]. A schematic drawing of the machine is shown in Fig. 4.

Limitations:

1. This machine is only used for dust sweeping.
2. Unable to pick up big types of garbage.
3. This machine can be operated by a toying tractor.
4. Large amount of power is required to operate this machine.
5. Not able to work on narrow areas.
6. Dust coming out while sweeping.

5. **Litter pickup vehicle**

Flemming [15] presented a garbage-picking mechanism that could be put on a moving vehicle. When the vehicle moves quickly, the waste becomes more and more entangled surface. There is an articulated arm in the mechanism. The arm has a waste pickup head at the end of it. The arm also has conveyor equipment to carry the picked-up litter that is installed on and made to fit the arm. The litter pickup head is made to move along the surface it is supporting while picking up any litter that comes into touch with it. The operator may search for and collect litter that is within arm's reach by fully controlling the head's position and operation from inside the vehicle's cab. The operation machine is represented schematically in Fig. 5.

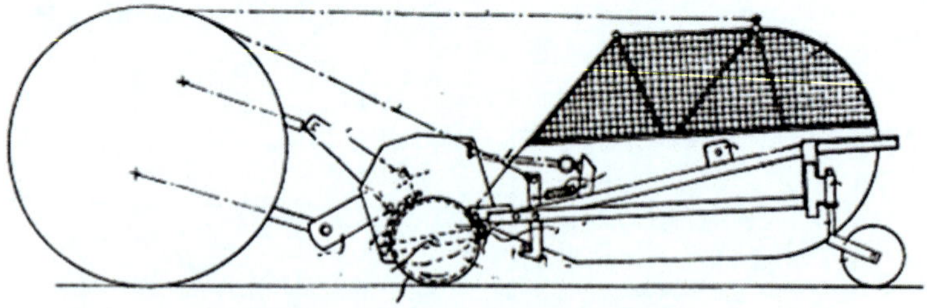

Fig. 4 Sweeping machine tractor attachment for sports grounds (Derek Burtt Hargreaves Adlington, near Macclesfield; Eric Stanforth, Macclesfield; Edward Hobbs, Poynton, all of, England, US3594844) [14]

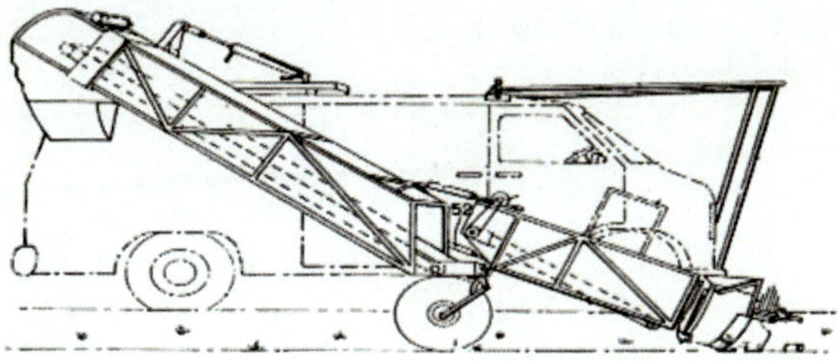

Fig. 5 Litter pickup machine (Jerry A. Flemming from Northeast Bellevue on 09 April 1975) [15]

Limitations:

The mechanism's limited waste pickup function is the litter pickup head. Because of this, very little trash was trapped. Litter within arm's reach may be sought out and collected by the operator. The load must be picked up using a vehicle and an arm. The machine has a smaller mechanical power advantage since moving requires more energy. Because a separate motor was needed to power the entire garbage picking system, it was pricey.

6. **Bottle and can picker including rotatable flexible loops**

Marsh and Johnson [16] showed off the next invention. Its structure is explained. The flexible loops come into sweeping contact with the ground as the carrier frame is moved over the surface, and the rotatable element is spun, dislodging the litter and pushing it upward and rearward into the waste container.

Working and Description:

A portable carrier frame, a rotatable member installed on the carrier frame, and several flexible loops attached to the rotatable member and extending radially are all components of a bottle and can picking equipment. There is a drive motor therefrom for collecting up the bottles, cans, and other garbage, and the rotatable member and a garbage container for collecting and holding the trash the loops have collected. The device consists of many circuits made of flexible materials such as padded vinyl or plastic and thin metal. These rings have two links and can be adjusted in length by loosening the screws, changing the size of the ring, and tightening the screws. These rings are designed to support the effect of collecting bottles and other garbage. The job of the loop is relative to the soil and ground to remove debris. Bottle and Can Picker is shown in Fig. 6.

Limitations of the Machine:

The machine employs several sets of flexible loops that are adjustable, but the loops are actually at a 90-degree angle from one location to the next. This means that the machine was designed to pick up larger objects, such as bottles and small cans; it is unable to pick up smaller items or waste other than bottle-sized waste. When a drum with attached flexible loops rotates, a power source motor is utilized. It operates with

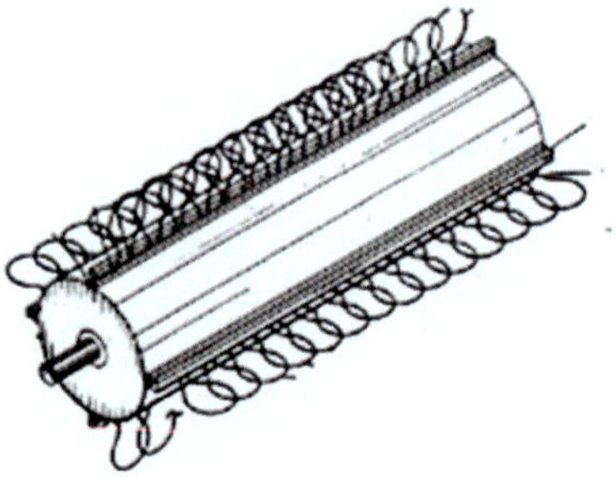

Fig. 6 A plurality of sets of flexible loops [16]

an electric motor, increasing the cost of maintenance, and it only collects large types of trash.

7. Manually operated path cleaner machine

A manually operated path-cleaning machine is the subject of study by Ramesh et al. [17]. This device has wheels with a cylindrical nylon brush 120 mm in diameter outside. 12 mm for the inner diameter, aluminum pulleys 52 mm is the size of the big pulley. The small pulley diameter is 26 mm, and the length of the V belt's center distance is 330 mm. V belt angle = 40, the inner diameter of a ball bearing is 12 mm. 32 mm for the outside diameter [17].

Working Description: A 26 mm-diameter aluminum tiny pulley is attached to a nylon cylindrical brush. The back wheel, which has an outside diameter of 120 mm, is attached to a 52 mm big pulley via a V-belt. The back wheel revolves with a larger-diameter aluminum pulley when the operator pushes the machine. Smaller pulley turns the nylon brush anticlockwise as the bigger pulley rotates in the other way. Anytime a nylon brush in a container is used to sweep up dust, the container has an inclined slope on the front side that collects the dust and other microscopic particles [17]. The design and fabrication of a manually powered path cleaner using pulley drives is shown in Fig. 7.

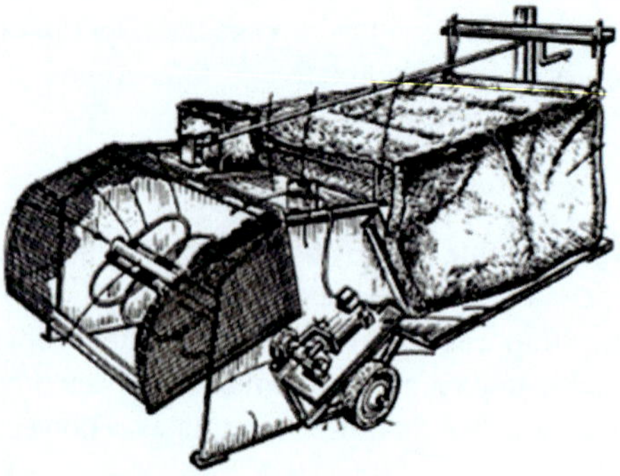

Fig. 7 Bottle and Can Picker Marsh. A. Johnson, New York on 16 February 1993 [16]

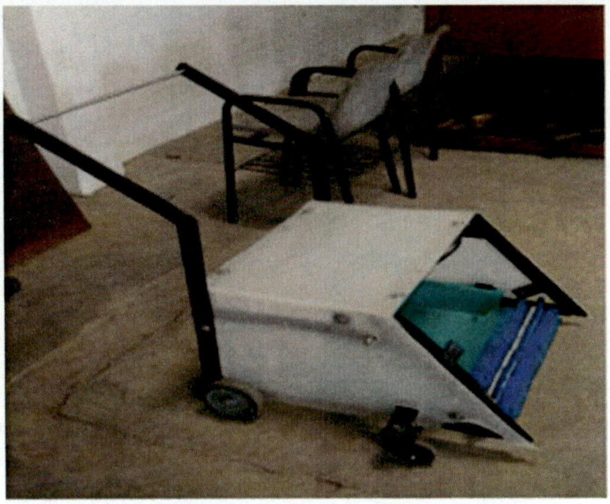

Fig. 8 Design and fabrication of manually powered path cleaner using pulley drives" international journal of engineering and advanced technology (IJEAT) Volume-9 Issue-1, October 2019 [16]

Limitations:

1. Only dust and small particles can be picked from this machine.
2. If any big type of garbage comes the machine will get stuck.
3. Less working area, the operator needs to run for longer time.
4. High energy is required for deep cleaning.
5. Less Storage Capacity.

3 Conclusion

Numerous methods for sweeping and gathering trash have been discussed and explained in this study. It also covered manual, electric, and gasoline-powered mechanisms. The future of sweeping has a very broad market to grab; all that remains is to remove the challenges and make garbage picking easier for operators to use and more cost-effective. The main consideration while constructing a machine is ergonomics. Whether or whether the operator uses it has a significant impact. By looking at all the patents that are mentioned in this paper, we can conclude that every machine has some limits that prevent us from using them, thus we create machines that transcend those limitations.

References

1. Aman khan, Anurag Pannase, Amol Sharnagat, Gaurav Gohane (2017) Study of multipurpose road cleaning machine. Int Res J Eng Technol 04(02)
2. www.iitmandi.ac.in
3. Chakraborty A, Bansal A (2013) Design of dust collector for rear wheel of four-wheeler. Int J Emerg Technol Adv Eng 3(7)

4. www.osha.europa.eu
5. Brijesh K, Karthik, Adarsh, Githin, Kevin Xavier (2019) Design and fabrication of waste collecting machine. Int Res J Eng Technol 06(05)
6. Amaljith, Kannan, Vaisakh, Vaishnav, Balakrishnan (2020) Automatic trash collector robot. Int J Sci Res Publ 10(4). 40 ISSN 2250–3153
7. National Center for Biotechnology Information. www.ncbi.nlm.nih.gov
8. www.worldsweepers.com
9. Ranjit Kumar M, Kapilan N (2015) Design and analysis of manually operated floor cleaning machine. Int J Eng Res Technol 4(04)
10. Adedeji KA, Adebesin E, Adekunle A (2020) Design and construction of a street sweeping machine. Int J Eng Appl Sci Technol 5(2). 2455–2143
11. Michaels A Manually operated sweeper.US Patent US3947912A. from 638 spruce st, philapedia, pa9106
12. Xiafeng X Road refuse sweeping vehical. Worldwide Patent CN203514238U, "项侠风"
13. Jia Z A kind of road that dust is dust removal integrated cleaning trolley. Worldwide Patent CN108755529A. 郑佳
14. Hargreaves DB Adlington, near Macclesfield, Eric Stanforth M, Edward Hobbs P, all of, England Sweeping machine tractor attachment for sports grounds. US Patent US3594844
15. Jerry A Flemming from northeast Bellevue, Litter pickup machine. US Patent, 09 April 1975
16. Johnson MA 16550 County Lind Rd., Newark, Ill. 60541, Bottle and can picker including rotatable flexble loops. US Patent
17. Ramesh P, Balaji DS, Gnanasekaran V, Gopi P (2019) Design and fabrication of manually powered path cleaner using pulley drives. Int J Eng Adv Technol (IJEAT) 9(1)

Chapter 9
Study of Leakage Currents in FinFET SRAM Cells

David Gavini, E. Pallavi, B. Kiran Kumar, and Pavankumar Bikki

Abstract Static Random Access Memory (SRAM) is an essential component in modern electronic devices, and its performance parameters need to be continually improved to keep up with advances in technology. However, the memory design with traditional CMOS has many issues in terms of density and leakage current. In this paper, we analyze a FinFET-based SRAM architecture that addresses leakage current problems. The various designs were simulated through the Cadence Virtuoso. The simulation results show that the FinFET design achieves a reduction in leakage current. FinFETs have a better gate control over the channel, which reduces threshold voltage variability, enabling better scalability. The design features a leakage controlling sleep transistor that significantly reduces leakage power. Moreover, FinFET designs has reduced subthreshold leakage, which results in lower standby power consumption. This is important for scaling because it enables the use of smaller devices, which are essential for integration the integraon density of SRAM arrays.

Keywords Complementary metal–oxide–semiconductor · Fin-shaped field-effect transistor · Leakage current · Power dissipation · Static random-access memory

1 Introduction

Static Random Access Memory (SRAM) is an important component in many electronic devices, including microprocessors, cache memories, and wireless communication devices. However, as technology continues to evolve, SRAM faces challenges such as power consumption, leakage current, and speed limitations. The FinFET (Fin-shaped Field-Effect Transistor) is a promising alternative to traditional MOSFETs (Metal–Oxide–Semiconductor Field-Effect Transistors) for SRAM designs due to its superior electrostatic control and reduced short-channel effects. Therefore, there

D. Gavini (✉) · E. Pallavi · B. Kiran Kumar · P. Bikki
Department of ECE, B V Raju Institute of Technology, Narsapur, India
e-mail: 19211a0448@bvrit.ac.in

V. Bindhu et al. (eds.), *Proceedings of Sixth International Conference on Inventive Material Science Applications*, Advances in Sustainability Science and Technology,
https://doi.org/10.1007/978-981-99-4189-6_9

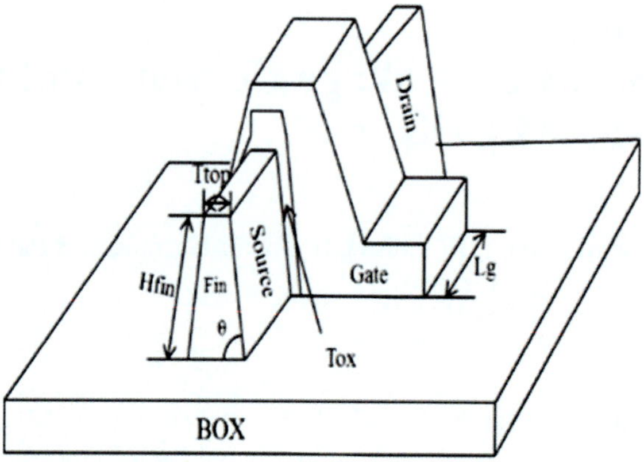

Fig. 1 Basic FinFET transistor

is a need to explore new designs and techniques for improving the performance of FinFET-based SRAM. Figure 1 shows the basic structure of FinFET [1].

The FinFET transistor has a fin-like structure that improves electrostatic control and reduces short-channel effects. FinFET SRAM offers several advantages over traditional SRAM, such as reduced power consumption, improved performance, and higher stability. However, FinFET SRAM also faces challenges, such as leakage current, which can significantly affect its performance [2].

As we previously indicated, the preceding graphic makes it very evident that FinFETs have a "fin"-like structure. Thus, Lg is the physical height of the silicon fin, and Hfin. Gate length of the FinFET, where Ttop is the silicon fin's thickness. In single-gate devices, the short-channel effect kicks in after the channel length reaches a certain point. If the channel length is then reduced after that, the device's performance starts to suffer because of the short channel effect. This is the single-gate device's primary flaw. Initially, double-gate FinFETs were introduced to successfully lessen the short channel effect and overcome it; however, they required two power supplies for each gate. Then, short-gate FinFETs were developed, which significantly reduced the short-channel effect only needed a single power source [3].

The main objectives of this paper are to propose a FinFET-based design that addresses leakage current issues and improves read and write functions, and to investigate the effectiveness of this design through simulation. Specifically, we propose a leakage-controlling sleep transistor inserted over a pull-up and pull-down network to significantly reduce leakage power. We also explore various techniques for improving reading and writing efficiency. The contributions of this paper are twofold; First, we present a novel FinFET-based SRAM design that addresses leakage current issues and improves performance. Second, we provide insights into the effectiveness of various performance optimization techniques for FinFET SRAM through simulation analysis [4, 5].

2 Literature Review

Several studies have explored the use of FinFET transistors in SRAM design. Monica et al. proposed a FINFET-based 8 T SRAM cell that showed improved read stability and reduced write voltage compared to traditional 6 T SRAM [6]. Gao et al. presented a FinFET-based SRAM design that demonstrated lower power consumption and improved stability compared to bulk-SI technology [7]. N Kuar et al. proposed a FinFET-based 10 T SRAM design that showed improved read and write performance and reduced leakage compared to traditional 6 T SRAM [8].

Leakage current is a major challenge in FinFET SRAM design. Several studies have analyzed the sources and impacts of leakage current in FinFET SRAM. For example, Kim et al. investigated the impact of gate-induced drain leakage (GIDL) on FinFET SRAM performance and proposed a method to mitigate this effect [9]. Yang et al. studied the impact of subthreshold current on FinFET SRAM performance and proposed a FinFET-based SRAM design that minimized subthreshold current [10].

Various techniques have been proposed to improve the performance of FinFET SRAM. One common approach is to insert a sleep transistor over the pull-up and pull-down networks to reduce leakage, which improves write stability and reduces write voltage. A sense-amplifier-based read-assist technique that improves read speed and reduces read power. Other techniques, such as adaptive body biasing and supply voltage scaling, have also been explored for FinFET SRAM performance improvement [11, 12].

An extra transistor is included in the 7 T SRAM cell for write operation than the standard 6 T cell, giving it a higher level of stability in noisy conditions. The extra transistor increases the SRAM cell's dependability and helps to prevent data loss. In Fig. 2, a 7 T SRAM cell with a FinFET is displayed [13].

The 9 T FinFET SRAM cells in Fig. 3 can aid in increasing stability and lowering power consumption. The reduced leakage current and better control over the flow of current offered by FinFETs can help to ensure that data is retained even in the presence of noise and other environmental factors. The additional pass gate transistor can also help to reduce power consumption by allowing for more efficient read and write operations. Overall, FinFET-based 9 T SRAM cells can be used for high-performance and low-power memory designs [14].

From Fig. 4, the FinFETs in 10 T SRAM cells provide several structural advantages that can lead to improved performance, lower power consumption, and smaller chip sizes, making it an attractive option for a wide range of applications [8].

A 12 T SRAM cell is shown in Fig. 5; there are three access transistors and nine storage transistors arranged in a cross-coupled configuration. The three access transistors are used to read and write data to the cell, while the nine storage transistors are used to store the data. The use of FINFETs in 12 T SRAM cells allows for better performance, lower power consumption, and smaller chip sizes. This makes FINFET-based 12 T SRAM cells ideal for use in a wide range of applications, including high-performance computing, mobile devices, and the Internet of Things (IoT) [15].

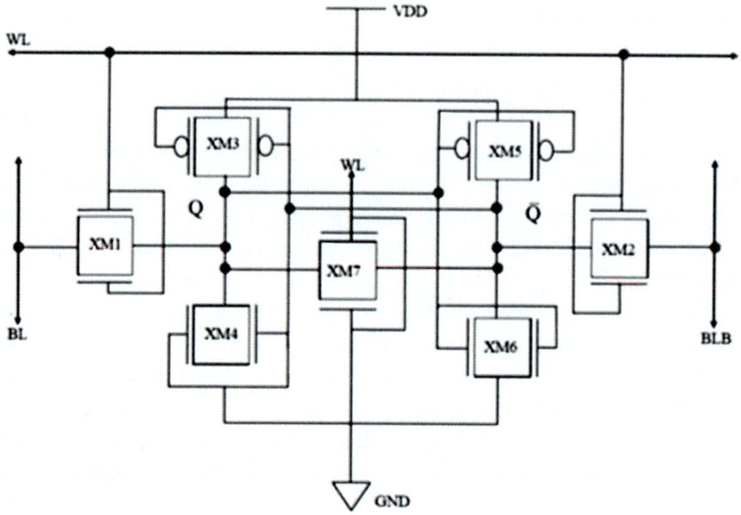

Fig. 2 7 T SRAM cell

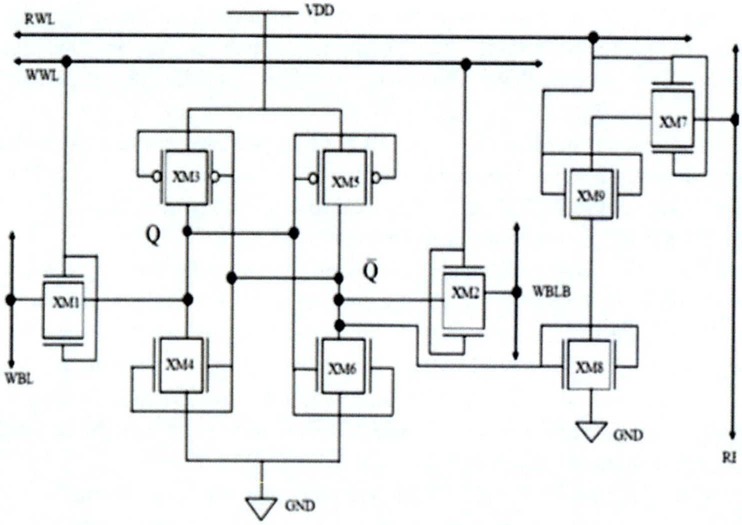

Fig. 3 9 T SRAM cell

In this paper, we build on prior work by proposing a novel FinFET-based SRAM design that addresses leakage current issues and improves read and write functions. We also investigate the effectiveness of various techniques for improving FinFET

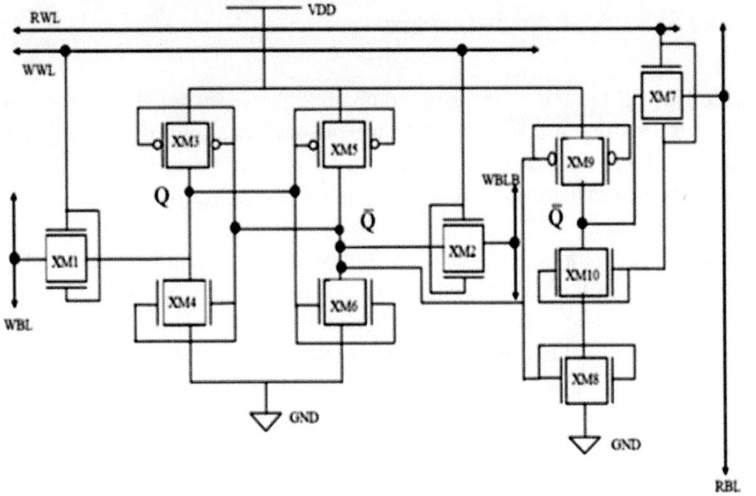

Fig. 4 10 T SRAM cell

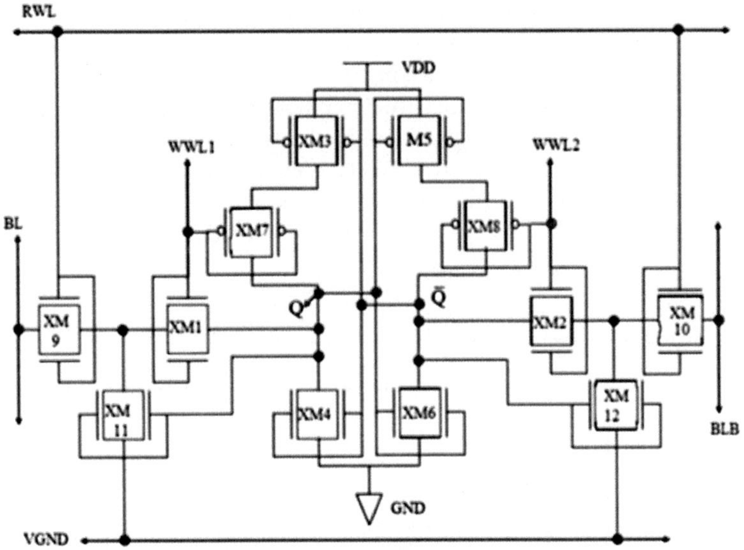

Fig. 5 12 T SRAM cell

SRAM performance, including the use of a leakage-controlling sleep transistor and other read-and-write optimization techniques. Our simulation-based analysis provides new insights into the potential for achieving improved efficiency in FinFET SRAM design.

3 FinEFT SRAM Archıtecture

The FinFET SRAM architecture is based on an 8 T cell design. The cell consists of four access transistors and four memory transistors, with two of each connected to the bit lines and the other two connected to the complementary bit lines. A leakage-controlling sleep transistor is inserted over the pull-up and pull-down networks to reduce leakage power.

To maintain the data integrity and usefulness of 6 T and 7 T SRAM cells, there are precise limits on the size of transistors. An 8 T SRAM, as shown in Fig. 6, is utilized to maintain read stability. The proportion of pull-down to access transistors describes the stability of 6 T and 7 T SRAM cells. A higher ratio results in improved data stability for 6 T and 7 T SRAM cells [11, 16]. A higher ratio results in improved data stability at the price of greater cell area and increased leakage power. An 8 T-cell solution was utilized to solve the issue of data storage destruction during read operations. With this implementation, the data output elements were read/write bits and word signal lines. The access transistors are designed with a gate length of 25 nm and a gate width of 180 nm, while the memory transistors have a gate length of 40 nm and a gate width of 120 nm. The FinFET SRAM cell is simulated and analyzed for its read and write functions, leakage current, power dissipation, and stability [17, 18]. Memories that need quick access times, little power consumption, and tolerance are implemented using SRAM cells shown in Fig. 7.

The leakage-controlling sleep transistor significantly reduces the leakage power, while the dynamic write assist technique improves write stability and reduces write voltage. Additionally, the FinFET SRAM cell achieves improved read and write functions, as demonstrated by simulation-based analysis. The use of FinFET technology also provides benefits such as improved scalability and reduced power consumption [19].

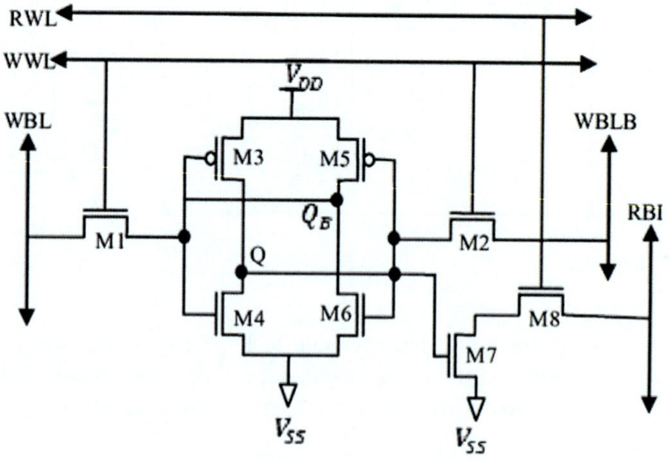

Fig. 6 8 T SRAM cell

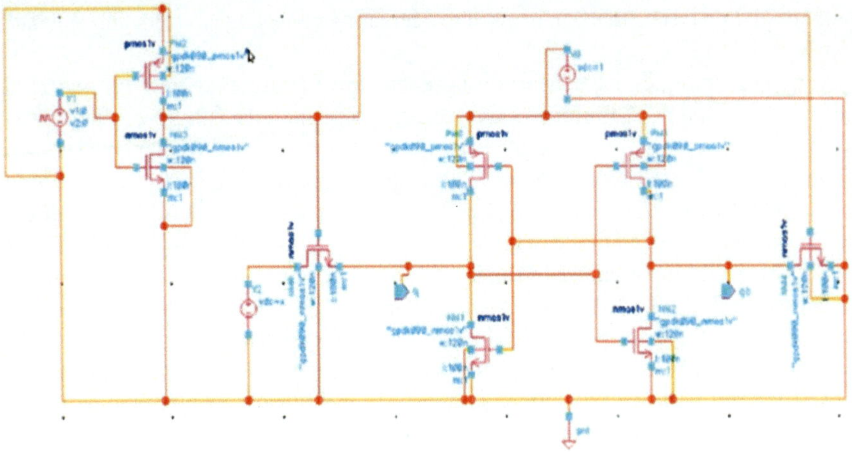

Fig. 7 Implementation of 8 T SRAM

4 Simulation Results Analyses

The simulation results were used to analyze the junction leakage currents, gate leakage currents, and sub-threshold currents of the FinFET SRAM cell. The analysis showed that the proposed design with the sleep transistors over the pull-up and pull-down networks significantly reduced the leakage power. Furthermore, the analysis revealed that the studied design resulted in a sub-threshold current reduction of 80% compared to the previous design. The reduction in sub-threshold current was achieved by optimizing the transistor size and using the sleep transistor.

The leakage current that passes through a transistor when it is operating in the subthreshold zone, when the gate voltage is lower than the threshold voltage, is known as the subthreshold current. Particularly for low-power applications, this current can have a considerable impact on the power consumption of integrated circuits. The subthreshold current is a crucial element in circuit design and optimization since it is greatly influenced by temperature and device characteristics like channel length and doping concentration. To decrease subthreshold current and increase power efficiency, a variety of methods can be used, including transistor stacking and threshold voltage modulation.

SRAM gate leakage current is a serious problem that may affect the memory cell's stability and dependability. Data flipping in an SRAM cell can result from gate leakage current, resulting in mistakes during data storage and retrieval. Low-power SRAM designs, where the operational voltage is decreased to reduce power consumption, worsen this issue by increasing gate leakage current. Designers can use a number of methods to alleviate this problem, including transistor stacking and body-biasing, to lessen the effect of gate leakage current on the SRAM cell. High-k gate dielectrics and metal gate electrodes are examples of sophisticated process methods and materials that can be used to decrease gate leakage current and enhance SRAM.

Various techniques have been proposed in the literature to improve the read and write efficiency of FinFET SRAM cells. These include voltage scaling, data-dependent write assist, and assist techniques such as column redundancy, block redundancy, and assist write. The write assist technique was implemented by adding an additional write driver circuit that provides extra voltage to the bit line when the data is written as "0". This technique improves the write efficiency of the SRAM cell. This results in faster write times and lower power consumption. The circuit design of the FinFET SRAM cell is implemented in Cadence Virtuoso using 32 nm FinFET technology at 25 °C and 110 °C with a load capacitance of 1 pF at a frequency of 100 MHz. The simulation results showed that the implementation of the data-dependent write assist technique improved the write efficiency of the proposed FinFET SRAM cell. The leakage current for 7 T SRAM is 5.38e−06, the leakage current for 8 T SRAM is 1.82e−05, the leakage current for 9 T SRAM is 1.82e−05, the leakage current for 10 T SRAM is 1.17e−06 and the leakage current for 12 T SRAM is 3.48e−06. The leakage current for the 7 T FinFET SRAM is 13.2e−09, 8 T FinFET SRAM is 12.82e−09, 9 T FinFET SRAM is 11.34e−09, 10 T FinFET SRAM is 10.43e−09 and 12 T FinFET SRAM is 8.11e−09. Overall, the FinFET SRAM architecture and circuit design, along with the implementation of the data-dependent write assist technique, result in a low-power SRAM cell with improved performance characteristics. These FinFET designs can serve as a reference for future FinFET SRAM designs. The V–I characteristics of the FinFET 8 T SRAM, 9 T SRAM, and 10 T SRAM designs are shown in Figs. 8 and 9.

The study of FinFET SRAM design was compared to traditional CMOS SRAM designs. The comparison showed that the FinFET design achieved significant improvements in terms of read and write delay and leakage power reduction. For instance, the FinFET SRAM cell achieved a read delay of 0.85 ns, while previous

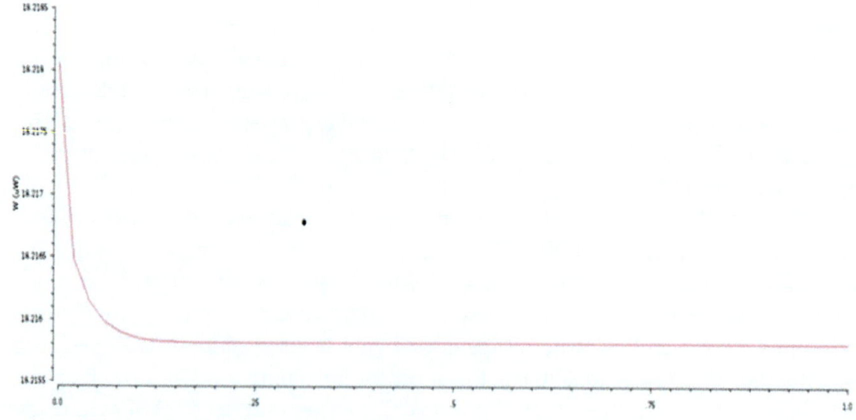

Fig. 8 V–I characteristics of FinFET 8 T SRAM

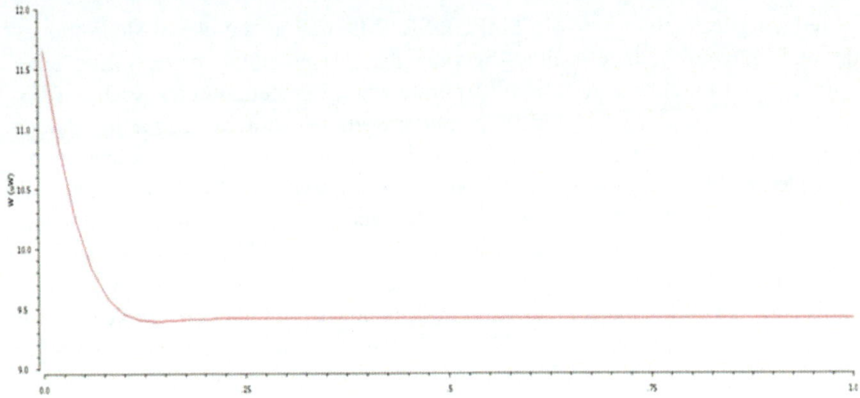

Fig. 9 V–I characteristics of FinFET 10 T SRAM

Table 1 Comparison of leakage current of various designs

Parameter	7 T	8 T	9 T	10 T	12 T
Supply voltage	1 V	1 V	1 V	1 V	1 V
Leakage power using CMOS	5.38E-06	1.82E-05	1.82-05	1.17E-06	3.48E-06
Leakage power using FinFET	13.2E-09	12.82E-09	11.34E-09	10.43E-09	8.11E-09

designs achieved read delays in the range of 1.0–1.5 ns. Similarly, the FinFET design achieved a leakage power of 8.11 nW, while CMOS designs had leakage power in the range of 3.48 uW. The simulation results demonstrate the effectiveness of the FinFET SRAM design and the performance optimization techniques employed. The design achieves improved performance in terms of read and write delay and reduced leakage power, making it a promising candidate for low-power SRAM applications. A comparison of leakage current of various designs is shown in Table 1.

5 Conclusions

In this paper, we studied various FinFET SRAM architecture that incorporates several techniques for reducing leakage currents. Through simulations using Cadence Virtuoso, we were able to analyze and optimize the power dissipation, leakage current, and performance of the proposed design. Our analysis showed that the FinFET designs significantly reduced the leakage current compared to CMOS designs, while also improving read and write efficiency. Specifically, FinFET 12 T SRAM design achieves a 199% reduction in leakage current compared to the 12 T SRAM CMOS design. Hence, there is room for further exploration of techniques for reducing leakage currents in FinFET SRAM, such as the use of sleep transistors and dynamic voltage scaling. Second, there is potential for exploring novel FinFET SRAM architectures that incorporate emerging technologies, such as resistive random-access memory (RRAM) and spin-transfer torque magnetic random-access memory (STT-MRAM). Finally, there is a need for further investigation into the tradeoffs between power dissipation, leakage current, and read and write efficiency in FinFET SRAM, as well as the impact of scaling on these parameters. Finally, this paper presents a promising design for FinFET SRAM that achieves significant improvements in performance and leakage current reduction. Future work in this area can build on this design and explore new techniques and architectures to continue pushing the boundaries of FinFET SRAM performance and efficiency.

References

1. Park J, Lee S, Jeong H (2023) Voltage boosted fail detecting circuit for selective write assist and cell current boosting for high-density low-power SRAM. IEEE Trans Circuits Syst I Regul Pap 70(2):797–805
2. Turi MA, Delgado-Frias JG (2020) Effective low leakage 6T and 8T FinFET SRAMs: using cells with reverse-biased FinFETs, near-threshold operation, and power gating. IEEE Trans Circ Syst II Express Briefs 67(4):765–769
3. Bikki P, Annapurna M, Vujwala S (2019) Analysis of low power SRAM design with leakage control techniques. In; 2019 TEQIP III sponsored international conference on microwave integrated circuits, photonics and wireless networks (IMICPW). IEEE
4. Raj B, Saxena AK, Dasgupta S (2011) Nanoscale FinFET based SRAM cell design: analysis of performance metric, process variation, underlapped FinFET, and temperature effect. IEEE Circuits Syst Mag 11(3):38–50
5. Bikki P, Annapurna M (2019) Analysis of low power novel SRAM design with self-biasing approach for ultra-low power applications. In: 2019 TEQIP III sponsored international conference on microwave integrated circuits, photonics and wireless networks (IMICPW). IEEE
6. Monica M, Chandramohan P (2016) Characterization of 8T SRAM cells using 16 nm FinFET technology. In: 2016 international conference on signal processing and communication (ICSC). IEEE
7. Gao R, Wang H, Xie Y (2015) Optimization and analysis of a FinFET-based 8T SRAM cell for low power applications. In: 2015 IEEE international conference on electron devices and solid-state circuits (EDSSC). Chengdu, pp. 1–2
8. Kaur N, Gupta N, Pahuja H, Singh B, Panday S (2016) Low power FinFET based 10T SRAM cell. In: 2016 Second international innovative applications of computational intelligence on power, energy and controls with their impact on humanity (CIPECH). Ghaziabad, India, pp. 227–233
9. Kim Y et al (2019) A 12T SRAM with improved read and write margins for low-power applications. IEEE Trans Circuits Syst II Express Briefs 66(6):1067–1071
10. Lakshmi TV, Kamaraju M (2021) A review on SRAM memory design using FinFET technology. Int J Syst Dyn Appl (IJSDA) 11.6:1–21
11. Li H, Zhao Y, Li X (2014) A low-power and high-speed 8T SRAM cell for FinFET technology. In: 2014 IEEE international conference on electron devices and solid-state circuits (EDSSC). Hong Kong, pp. 1–2
12. Alioto M (2015) Ultra-low voltage, ultra-low power FinFET SRAM design. In: Proceedings of IEEE International Symposium on Circuits and Systems (ISCAS), pp. 1557–1560

13. Sneha G, Krishna BH, Kumar CA (2017) Design of 7T FinFET based SRAM cell design for nanometer regime. In: 2017 International conference on inventive systems and control (ICISC). Coimbatore, India, pp. 1–4
14. Yang Y, Park J, Song SC, Wang J, Yeap G, Jung S-O (2015) SingleEnded 9T SRAM cell for near-threshold voltage operation with enhanced read performance in 22-nm FinFET technology. IEEE Trans Very Large Scale Int (VLSI) Syst 23(11):2748–2752.
15. Karamimanesh M et al (2021) A robust and write bit-line free sub-threshold 12T-SRAM for ultra low power applications in 14 nm FinFET technology. Microelectron J 118: 105185
16. Kumar TS, Tripathi SL (2022) Comprehensive analysis of 7T SRAM cell architectures with 18nm FinFET for low power bio-medical applications. Silicon 14.10:5213–5224
17. Yang JP, Song SC, Wang J, Yeap G, Jung S-O (2015) Single ended 9T SRAM cell for near-threshold voltage operation with enhanced read performance in 22-nm FinFET technology. IEEE Trans Very Large Scale Int (VLSI) Syst 23(11):2748–2752
18. Pal S, Bhattacharya A, Islam A (2014) Comparative study of CMOS-and FinFET-based 10T SRAM cell in subthreshold regime. In: 2014 IEEE international conference on advanced communications, control and computing technologies. IEEE
19. Raushan RK et al (2021) Implementation of 12T and 14T SRAM bit cell using FinFET with optimized parameters. Trans Electr Electron Mater 22:328–334

Chapter 10
Advanced Characterization of Estolides from Coconut, Sunflower and Karanja Oils

P. V. Maneesh Kumar and T. Jagadeesha

Abstract Cold flow properties of vegetable oils are generally subpar to industrial lubricants. This is one major drawback that limits the usage of otherwise abundant vegetable oil. Estolides belong to the class of esters synthesis from naturally occurring oils as carboxylic acid functionality of a fatty acid reacts, forming ester linkage on an unsaturated site of another fatty acid. Estolide number tells the mean number of fatty acids attached to the base fatty acid that varies with respect to synthesis temperature, reacting saturated fatty acids, and other parameters. Though vegetable oils show excellent biodegradability, their cold flow properties are poor compared to functional lubricants. In this present study, three vegetable oils have been synthesized to obtain their estolides, they are coconut, sunflower, and karanja oils to analyze the variations. Lubricating machines with mineral-based lubricants have become a major cause of pollution now. To use natural oils for making lubricants instead of petroleum is something mainstream researchers have been looking at for a few decades. Though vegetable oils show excellent biodegradability, their cold flow properties are poor compared to functional lubricants. Biodegradable lubricant oils must have excellent cold flow properties as well as competitive costs in order to be accepted widely as a base stock for industrial lubricants.

Keywords Estolides · SPARTAN · Saturated fatty acids · Pour point

1 Introduction

Vegetable oil belongs to the family of fats and lipids. Coconut oil is an abundant natural oil. It is mentioned among the lubricants used in the pre-mineral era. Coconut oils show good lubricating properties in many aspects like good viscosity index, fine lubrication, above average flash point, and lower evaporative losses [1, 2]. A major

P. V. Maneesh Kumar (✉) · T. Jagadeesha
Department of Mechanical Engineering, National Institute of Technology, Kerala, 673 601 Calicut, India
e-mail: maneesh_p210081me@nitc.ac.in

V. Bindhu et al. (eds.), *Proceedings of Sixth International Conference on Inventive Material Science Applications*, Advances in Sustainability Science and Technology,
https://doi.org/10.1007/978-981-99-4189-6_10

demerit of coconut oil is that it exhibits very poor cold-flow properties. It is due to the predominance of saturated fatty acids that are present. The pour point of coconut is 23 °C. Significant reduction in pour point is essential for considering coconut oil as base stock of lubricants in cold climatic and temperate conditions. Sunflower oil is another abundant oil that can be utilized. Sunflower oil is highly unsaturated in nature. Sunflower oil exhibits better cold-flow properties than coconut oil (−17 °C), but it also needs improvement in comparison with commercial lubricants. Karanja oil is also an abundantly available natural oil with a pour point in the range of 4 °C. Karanja oil contains fatty acids of saturation and unsaturation [3]. Synthesizing estolides has proven to be improving the cold flow properties of natural oils [3].

Esterified oligomers are further reacted to 2-ethylhexanol to increase molecular weight, the degree of oligomerization results in a better pour point [4]. The fatty acids attached to the double bonds cause the formation of estolides resulting in better cold flow characters. Generally, plant base oils exhibit pour points in the range of around 10 to 20 °C. Surprisingly, estolides clot at 5 °C to 45 °C that are comparable with commercial lubricants, viscosities are also improved with oligomerization [5]. Estolide number is an indicator of the physical properties of the synthesized estolides. The characteristics can be further improved by capping the estolides with 2-ethyl hexanol for additional improvements [5].

A lubricant's main purpose is to reduce the friction acting between the mating surfaces that causes deformation, deterioration, and stress induction. For this, any material that can act as a protective layer between the surfaces that can also reduce the interlocking of mating surfaces causing stress buildup can be employed. An ideal lubricant should have good viscosity to maintain a strong film, good viscosity index, good flash point to withstand heat in mating surfaces, and low volatility along with other properties. Based on the operating conditions, a lubricant is selected for application, in cold working conditions, only a lubricant with very good pourpoint can be used otherwise, the lubricant will solidify causing all possible problems. A refrigeration lubricant must be thermally stable over a wide range of temperatures and must be chemically inert too. This work is an attempt to produce a good lubricant from coconut oil by eliminating one of its major defects, which is the pour point of around 20 °C that makes it obsolete in cold conditions.

2 Experimental Procedure

2.1 Materials

Coconut, sunflower, karanja, and castor oils were procured locally as they are readily available for cooking purposes. Oleic acid, perchloric acid, 2-ethyl hexanol, KOH, ethanol, and silica gel were obtained from SIGMA ALDRICH Chemical Company, supplied by Alpha Chemicals Ernakulam, Kerala.

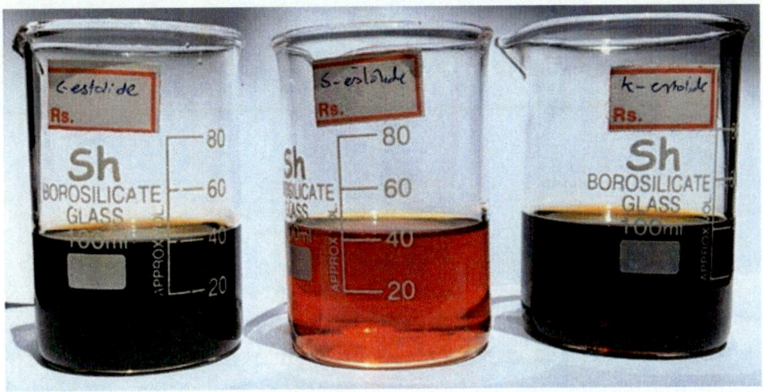

Fig. 1 Estolides synthesized from coconut, sunflower, and karanja oils

2.2 *Preparation of Estolides*

Estolides were synthesized from coconut, sunflower, and karanja oils through condensation reactions that are catalyzed by acids. The synthesis is in a 750 ml 3-necked flask at atmospheric pressure. The reactant mixture for estolide contains 90 g base oil, 10 g of castor oil with 20.3 g of oleic acid. The mixture was heated in a Rota mantle to a temperature of 60 ± 0.1 °C. A magnetic stirrer was used for mixing. When the required temperature was attained, 3 g of perchloric acid ($HClO_4$) was added drop by drop. Then the reactants were heated for 24 h at atmospheric pressure for obtaining estolides. After 24 h, 56.5 g of 2-ethyl hexyl alcohol was added to the reactant flask and the reactants were further heated for 2 h to get estolide esters. For quenching, 2 g potassium hydroxide (KOH) mixed through ultrasonic stirring for 180 s in 90% ethanol–water mixture was added. The completed reactions were cooled for an additional 30 min. The reactant mixture was transferred to a separating funnel and 100 g hexane was added for extracting the organic layer. The estolide synthesized was then further water-washed with warm water at 60 °C till the pH of washed water obtained equals the pH of normal water. For further water removal, the estolides were kept in an open beaker surrounded by silica gel in a closed vessel for 24 h. Commercially available coconut, sunflower, and karanja oils were used for experiments (as in Fig. 1).

2.3 *FTIR Analysis*

Fourier-transform infrared spectroscopy helps to understand the infrared spectrum of the emission and absorption of matter. An FTIR spectrometer deals with high-resolution spectral data across a vast range of spectra. This mechanism gives better results than a dispersive spectrometer, which only measures the intensity over a small spectra of wavelengths at one point of time. We can confirm estolide synthesis if the results from FTIR show peaks at desired points.

2.4 Tribological Study

Tribological studies were conducted on a DUCOM 4 ball tester as per ASTM D 4172-B standards with a load of 392 N, temperature of 75 °C, and speed of 1200 RPM for 3600 s. The obtained friction coefficient values and wear scar diameters are analyzed to evaluate the improvements in estolides synthesized over base oils. The variation of frictional characteristics was analyzed between base fluids and their estolides for comparison.

2.5 Viscosity Analysis

The viscosity of samples was analyzed on a BROOKFIELD viscometer as per ASTM D 2983 standards. The samples were analyzed at constant torque and varying speed rates on rotating a spindle immersed in an oil sample between 50 and 200 RPM in temperature ranging from 40 to 100 °C at 10 °C intervals. The value of kinematic viscosities at 40 and 100 °C are obtained from the viscometer. The viscosity index was identified as per ASTM D2270 to compare with commercial lubricants.

3 Results and Discussion

3.1 FTIR Analysis

FTIR analysis is a spectroscopy in which infrared rays are passed through the sample to evaluate the properties of the sample being tested. Based on the absorption of certain wavelengths by the sample, now we are able to characterize the bonds present in the sample. From the samples tested, the following results were obtained. The peaks at 2922/cm are C–H bonds representing fatty acid structure. The peaks at 1734/cm are representing ester functional groups (2-ethylhexyl ester). The peaks at 1041/cm are indicating C = O bond. From this, we can confirm that the synthesis of estolides is successful. FTIR analysis of Coconut, Sunflower, and Karanja estolides are shown in Figs. 2, 3 and 4. DSC Analysis of estolides of Coconut, Sunflower, and Karanja is shown in Fig. 5.

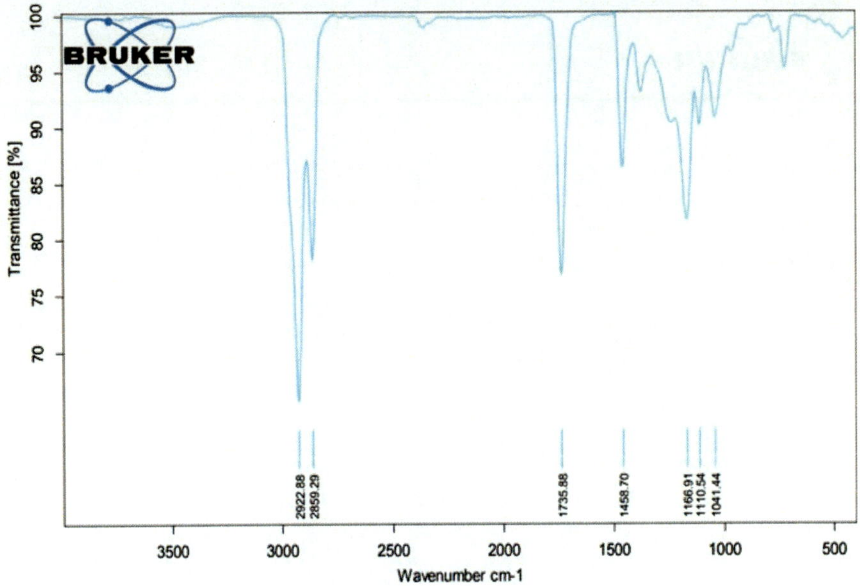

Fig. 2 FTIR of coconut estolide

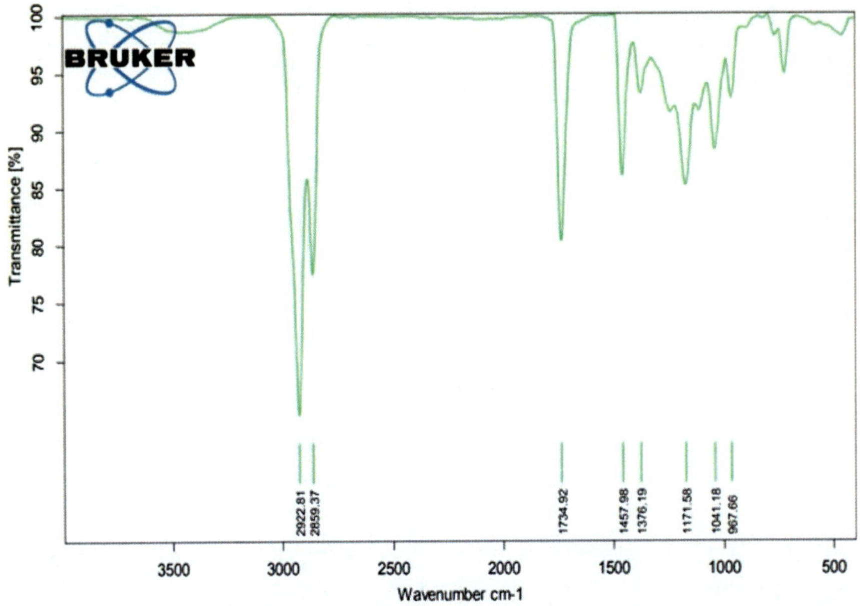

Fig. 3 FTIR of sunflower estolide

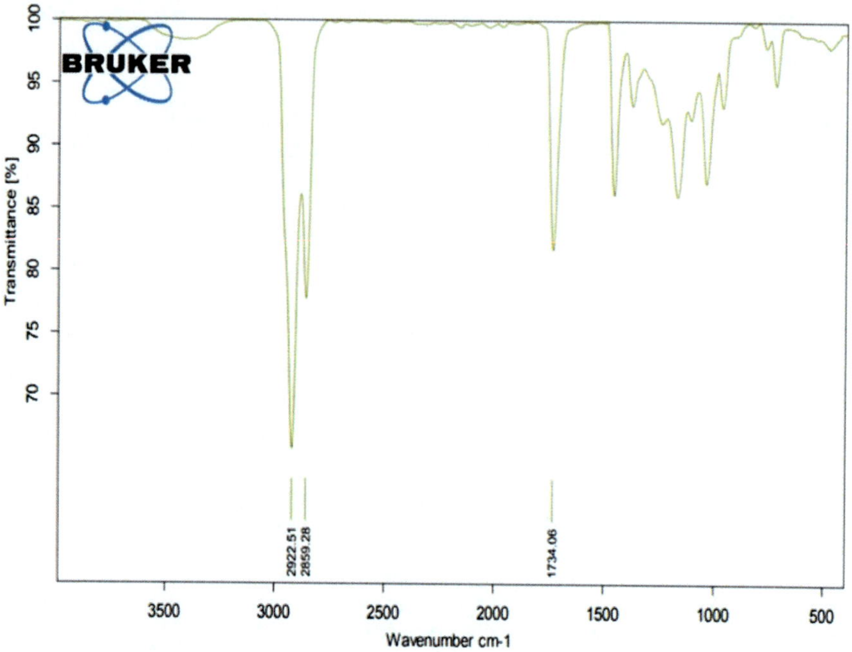

Fig. 4 FTIR of karanja estolide

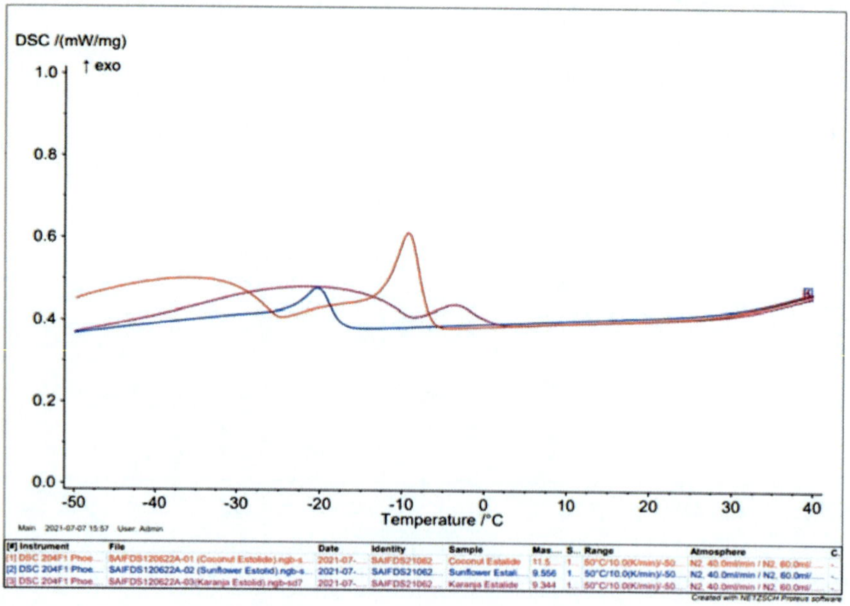

Fig. 5 DSC analysis of estolides of coconut, sunflower, and karanja

3.2 *Pour Point Analysis—DSC*

Pour point is one major aspect of lubricants and is one area where natural oils show poor properties. Pour point is also the major demerit of coconut oil as a lubricant because it will solidify at a temperature as low as 22 °C, which makes it unusable for lubricant applications even though it has many other merits. Though pour point could have been evaluated from the pour point apparatus, for more precise results the samples were sent for DSC. The samples were sent to STIC (Sophisticated Testing and Instrumentation Center, Cochin) for DSC—Differential Scanning Calorimetry testing to determine their crystallization parameters. Cloud points are determined from the college laboratory, pour point values are obtained from DSC. Oligomers with branching have complex structures resulting in lower Pour points [6]. Coconut oil shows a significant decrease in pour point from 22 °C to −3 °C. Sunflower oil's pour point decrease from −18 °C to −21 °C. Karanja oil's pour point decreased from 9 °C to −12 °C as per the ASTM D97 test and according to DSC thermograms, the pour points have been lowered to −9 °C, −20 °C and −3 °C for coconut, sunflower and karanja estolides, receptively. Estolides are known for their improved cold-flow properties, and the results also state the same.

3.3 *Viscosity Analysis*

Viscosity can be considered the major property of lubricating oils as it is the property that reduces the friction acting upon the mating surfaces. Most natural oils have good viscosity characteristics due to the presence of large molecules in them. Here, the variation of viscosity between natural oils and their estolide samples is shown in the graph (Fig. 6). Coconut estolide is showing reduced viscosity compared to pure coconut oil. Sunflower estolide shows improved viscosity at 40 °C but is showing reduced viscosity at 100 °C. Karanja estolide shows reduced viscosity compared to karanja oil. It is to be inferred that estolides show poor Viscometric properties compared to their base oil properties. The viscosity variations of coconut oil are from 25 cP at 40 °C to 8.4 cP when the temperature is 100 °C with a viscosity index of 152. Whereas the estolide shows a viscosity of 11.2 cP when the temperature is 40 °C to 4.07 cP at 100 °C with a VI value of 224. The viscosity variations of sunflower oil are from 35.9 cP when the temperature is 40 °C to 15.566 cP at 100 °C with a VI of 167, whereas the viscosity of estolides is 53.6 cP at 40 °C to 13.96 cP at 100 °C with a VI value of 130. The variations in karanja oil are from 46.5 cP at 40 °C to 14.2 cP at 100 °C with a VI of 101, and the estolides are showing a viscosity of 22.2 cP at 40 °C to 13.96 cP at 100 °C with a VI of 134. Variation of VI of oils, and their estolides compared with SAE20W40 are tabulated in Table 1.

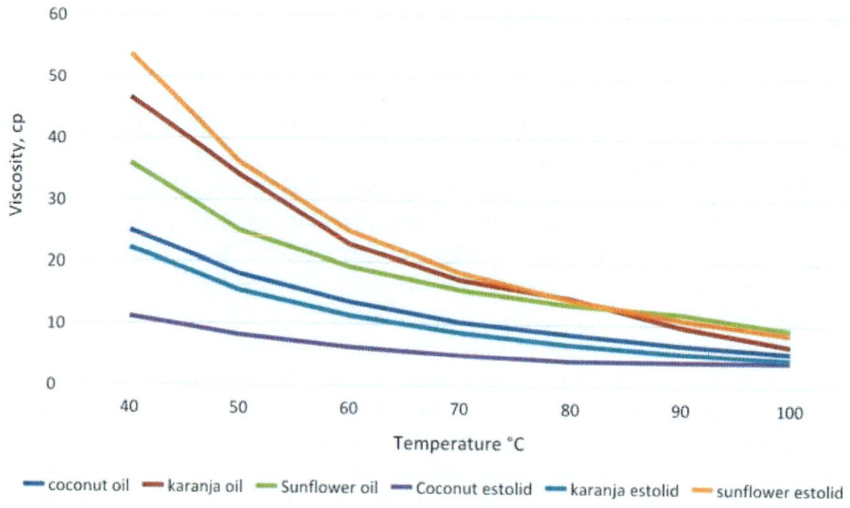

Fig. 6 Viscosity variation of base oils and their estolides

Table 1 Variation of VI of Oils, their estolides compared with SAE20W40

Sl no	Sample	Viscosity index
1	Coconut oil	152
2	Coconut estolide	224
3	Sunflower oil	167
4	Sunflower estolide	130
5	Karanja oil	101
6	Karanja estolide	134
7	SAE 20W40	125

3.4 Coefficient of Friction

Friction characters of the base oils and their estolides were analyzed on a Ducom 4 ball tester to evaluate their coefficient of friction and these balls were used for the examination of the wear scar diameters to evaluate properties and to compare with their base oils (Figs. 7, 8, 9, 10, 11 and 12).

From Table 2, coconut estolide shows a reduction in the coefficient of friction over the base oil. Sunflower estolide shows increased coefficient of friction over its base oil. Karanja estolide shows an increase in the coefficient of friction over its base oil. The wear scar diameter of coconut estolide has increased from 591.48 to 600.78 μm. The wear scar diameter of sunflower estolide has increased from 723 to 745 μm. The scar diameter of karanja oil is reduced from 524 to 544 μm. Coefficients of friction and wear scar diameter show insignificant variations between the base oil and estolide forms.

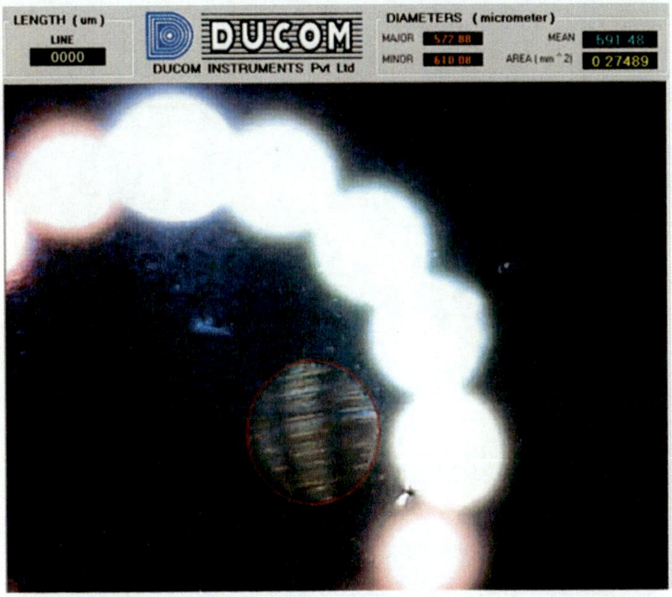

Fig. 7 Wear scar using coconut oil

Fig. 8 Wear scar using coconut estolide

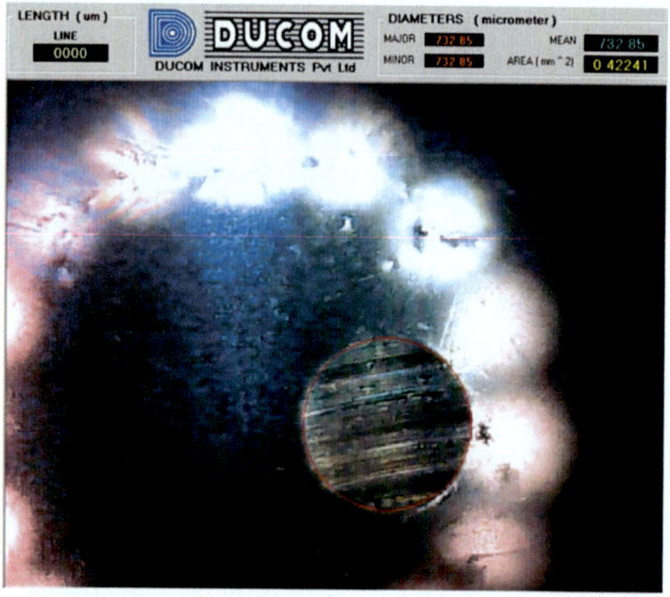

Fig. 9 Wear scar using sunflower oil

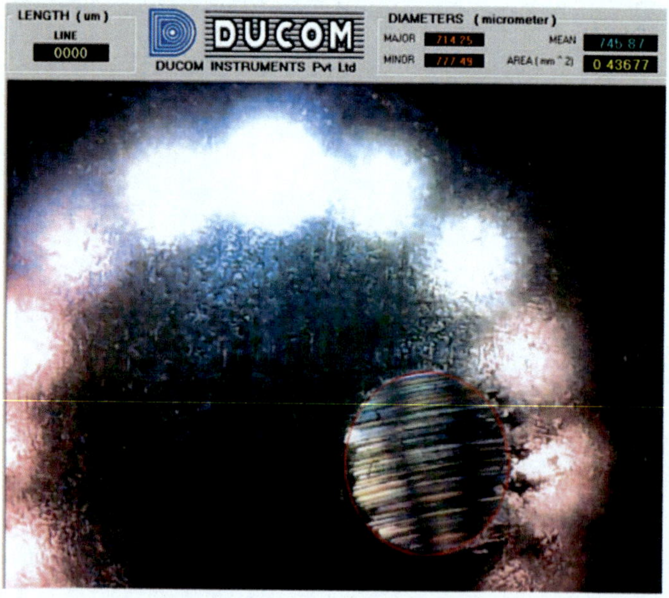

Fig. 10 Wear scar using sunflower estolide

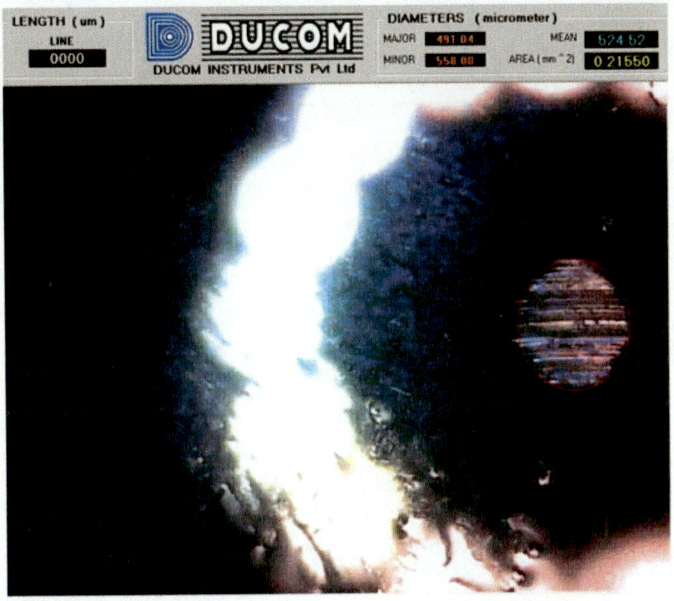

Fig. 11 Wear scar using karanja oil

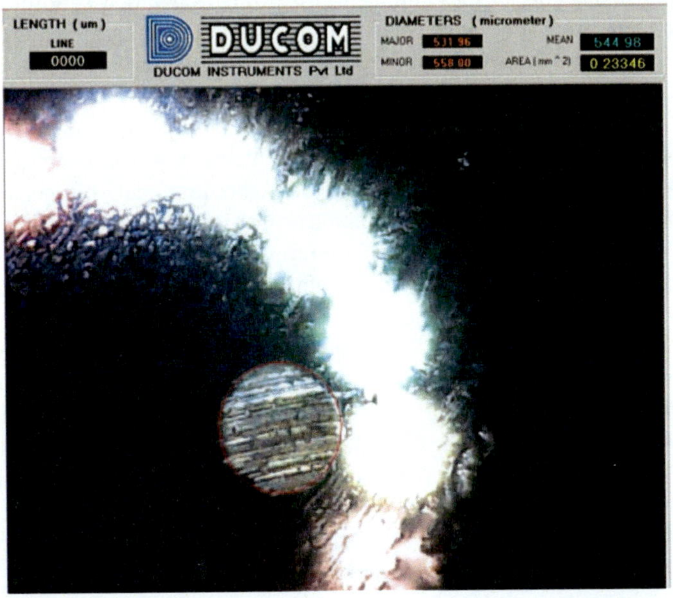

Fig. 12 Wear scar using karanja estolide

Table 2 Variation of CoF and WSD of oils and their estolides

Sl no	Sample	COF	WSD
1	Coconut oil	0.0779	591.48
2	Coconut estolide	0.0509	600.78
3	Sunflower oil	0.0743	732.85
4	Sunflower estolide	0.0825	745.67
5	Karanja oil	0.0530	524.52
6	Karanja estolide	0.0702	544.8

4 Conclusion

Though estolides have insignificant improvements over their base oils in terms of viscosity and coefficient of friction, their cold-flow properties show significant improvements. Vegetable oils make macro-crystalline figures through ordered arrangements of triacylglycerol, with branching at possible unsaturation sites like oligomerization and alkoxylation can improve the cold flow characters [3]. Pour point, which is a major demerit of vegetable oils, can be improved to a lower value with branching at possible locations on unsaturated sites with chemical modification. As the estolides are formed, the branching at the unsaturated sites leads to a slower uniform stacking of these molecules, slowing down their solidification [6]. The alcohol part from the ester attached also helps in reducing the pour-point. So the synthesized estolides were further converted to 2-ethylhexyl estolide esters by reacting them with 2- ethylhexanol for an additional 2 h for a further lower pour-point. Nature has an abundant source of fatty acids that can satisfy the raw material requirements of most needs [7]. Recent developments in this area are confirming the possibilities of natural industrial oils [8]. Another promising bio-derived lubricants are polyols showcasing good lubrication properties [9]. Adding more molecular weight to component molecules can help improve properties further [10]. Estolide esters can also be converted into biofuel rather than bio-lubricants is another ray of hope [11]. The cold-flow characteristics of thus produced biofuels can also be improved to industrial standards [12]. There are many methods to improvise the cold flow characteristics of the fluids to industrial requirements [13]. Polymers and vegetable oil-based composites are replacing major industrial fluids in most applications [14]. Another advantage of estolides is that all synthesis processes happen at temperatures as low as 60 °C, which means the cost of production of samples is low compared to other processes. Further improvements are to be made to make it a proper industrial-grade lubricant. Blending with other oils, addition of nanomaterial, etc. are some methods to improve the characteristics.

References

1. Isbell TA, Edgcomb MR, Lowery BA (2001) Physical properties of estolides and their ester derivatives. Ind Crops Prod 13:11–20
2. Sajeeb A, Rajendrakumar PK (2019) Comparative evaluation of lubricant properties of biodegradable blend of coconut and mustard oil. J Clean Produc 240:118255
3. Ajithkumar G (2009) Analysis of the pour point of coconut oil as a lubricant base stock using differential scanning calorimetry. Lubric Sci 01
4. Ho CK, McAuley KB, Peppley BA (2019) Biolubricants through renewable hydrocarbons: A perspective for new opportunities. Renew Sustain Energy Rev
5. Cermak SC, Isbell TA (2002) Physical properties of saturated estolides and their 2-ethylhexyl esters. Indust Crops Produc
6. Greco-Duarte J, Collaço ACA, Costa AMM, Silva LO, Da Silva JAC, Torres AG, Fernandez-Lafuente R, Freire DMG (2019) Understanding the degree of estolide enzymatic polymerization and the effects on its lubricant properties. Fuel
7. Ray AK, Stark RE (1998) Isolation and molecular structure of an oligomer produced enzymatically from the cuticle of lime fruit, phytochemistry, vol 48, no 8, pp 1313–1320
8. Perera M, Yan J, Xu L, Yang M, Yan Y (2022) Bioprocess development for biolubricant production using non-edible oils, agro-industrial byproducts and wastes. J Clean Produc 357:131956
9. Åkermana CO, Gaber Y, Ghani NA, Lämsä M, Hatti-Kaul R (2011) Clean synthesis of biolubricants for low temperature applications using heterogeneous catalysts. J Molec Catalys B Enzymatic 72:263–269
10. Bantchev GB, Cermak SC (2022) Correlating viscosity of 2-ethylhexyl oleic estolide esters to their molecular weight. Fuel 309:122190
11. de Haro JC, del Prado Garrido M, Perez A, Carmona M, Rodríguez JF (2018) Full conversion of oleic acid to estolides esters, biodiesel and choline carboxylates in three easy steps. J Clean Produc 184:579e585
12. Smith PC, Ngothai Y, Nguyen QD, O'Neill BK (2010) Improving the low-temperature properties of biodiesel: methods and consequences. Renew Energy 35:1145–1151
13. Sierra-Cantora JF, Guerrero-Fajardo CA (2017) Methods for improving the cold flow properties of biodiesel with high saturated fatty acids content: a review. Renew Sustain Energy Rev 72:774–790
14. Zhang C, Garrison TF, Madbouly SA, Kessler MR (2017) Recent advances in vegetable oil-based polymers and their composites. Progress Polym Sci 71:91–143

Chapter 11
Manufacturing of $AlSi_{10}Cu_2$ Foam Using Centrifugal Casting Route and $LiAlH_4$ Foaming Agent

Anandarao S. Kashid and Abhijit A. Patil

Abstract Foams are intentionally produced porous lightweight materials useful in structural, mechanical, acoustic, thermal, and domestic applications. Aluminum and its alloy foams are gaining importance due to lower melting temperatures, lower densities, and ease of manufacturing by a variety of techniques, light in weight, good strength-to-weight ratio, good anticorrosive properties, and aesthetic in appearance. Powder metallurgy route of foam manufacturing is used to check the suitability of hydrides like $LiAlH_4$ as foaming agents. LM2 (AlSi10Cu2) alloy is cold compacted at 200 MPa for 5 s, and at 6000C hydrogen release is observed for a minimum 5% weight of lithium containing blowing agents. This is an alternative for TiH2 foaming agent by powder metallurgy route. The foam production can be analyzed by visual observation, microstructure observation and nondestructive testing methods like X-ray diffraction analysis. A new approach of centrifugally cast powder metallurgy precursor method is introduced in this paper which can generate 45 to 60% porosity for batch production requirements at low cost.

Keywords Foaming agents · Foaming process · Metal foam · Powder metallurgy · Centrifugal casting

1 Introduction

Material Science is gaining the attention of all branches of engineering as a tool of development in the product and manufacturing areas. The development of composites, nanomaterials, and on-demand composites are satisfying complex property

A. S. Kashid (✉)
Department of Mechanical Engineering, N K Orchid College of Engineering & Technology, Solapur, Maharashtra, India
e-mail: askashid@gmail.com

A. A. Patil
Department of Mechanical Engineering, S. G. University, Kolhapur, Maharashtra, India
e-mail: abhijit.patil@sanjayghodawatuniversity.ac.in

V. Bindhu et al. (eds.), *Proceedings of Sixth International Conference on Inventive Material Science Applications*, Advances in Sustainability Science and Technology,
https://doi.org/10.1007/978-981-99-4189-6_11

requirements of design and production streams. Foams and cellular materials are basically intentionally produced porous materials for lightweight applications of metals, plastics, and ceramic materials family. The porous products are manufactured conventionally by powder metallurgy process with required quality and property combinations. But the cost of manufacturing is high as well as it is suitable for mass-scale production only. Metal foams are an important group of materials having applications in structural, mechanical, acoustic, thermal, and domestic applications. Aluminum and its alloy foams are gaining importance due to lower melting temperatures, lower densities, and ease of manufacturing by a variety of techniques [1]. Aluminum foams are light in weight, good strength-to-weight ratio, good anti-corrosive properties, and are aesthetic in appearance. Open-grain and close-grain aluminum foams are mostly used in automotive, aerospace, building and construction, healthcare, and energy industries [1]. A centrifugally cast powder metallurgy precursor in a stationery furnace is a new approach. It combines centrifugal casting and powder metallurgy precursor methods of foam manufacturing.

The paper is divided into five sections. Section 2 is related to literature review of methods available for foam manufacturing. Section 3 gives details of the experimental procedure and observations. Conclusion of the experiment is given in Sect. 4

2 Literature Review

Currently, there are six methods of use for commercial manufacturing of metal foam.

- **Bubbling gas in molten alloy** [2, 3]: In this method, foams are made by injecting gas bubbles into molten aluminum. It can produce highly porous foam having a large pore size. It is a low-cost method as compared with the power compact and Alporas method.
- **Stirring of foaming agent in molten alloy** [2, 4]: It is also called as 'Alporas' method. Metal powders and blowing agents are mixed at 6800C to 7200C and stirred for 6 min to increase viscosity. Due to the decomposition of blowing agent, gas gets released, which causes the foaming process.
- **Consolidation of metal powder with particulate foam powder** [5, 6]: To produce aluminum foam, aluminum powder or alloys of aluminum are mixed with foam agents and compacted. The foaming agent gets uniformly distributed in the foam obtained. The density of metal foam depends on foaming parameters and agents.
- **Vapor phase deposition of metal on polymer foam precursor** [7, 8]: Foaming take place by galvanically deposing polymer in ionic state of metal. Ions of metal and vapors are deposited on polymer and then the polymer is removed to obtain the foam.

- **Sintering of metal powder with a leachable foam powder** [9, 10]: In manufacturing of metal foams using sintering, thermal treatment is used to entrap the gas in the first step, and it is expanded in the second step. It forms uniform porous metal due to creep expansion.
- **Hydrogen gas dissolution in hot liquid metal under pressure** [11–13]. A pore-free melt having completely solved hydrogen is prepared at high pressure and pressure. Now, two-phase region solid and gas is formed by reducing temperature and pressure. The gas gets accumulated in the form of fine bubbles and forms foam if the cooling rate and pressure profile are selected correctly.

From the literature survey, it is observed that there is a scope for combining two of the existing manufacturing processes and to develop a new technique.

3 Experimental Setup

A small, batch production machine is developed for designing and testing for aluminum and its alloys as shown in Fig. 1.

A machine is fabricated, consisting of a rotating vertical mechanism for metallic die of mild steel mounted on a disc. This setup is fitted in an electrical resistance furnace having a maximum temperature of 800 °C having a heating rate of 11.5 °C / min [4]. The die is rotated by the electric motor at 1440 rpm. The total cycle time for the 250 gm batch is 50–60 min. The variables like rotational speed, weight of foaming agent, temperature of melting can be studied on this setup. The chemical analysis

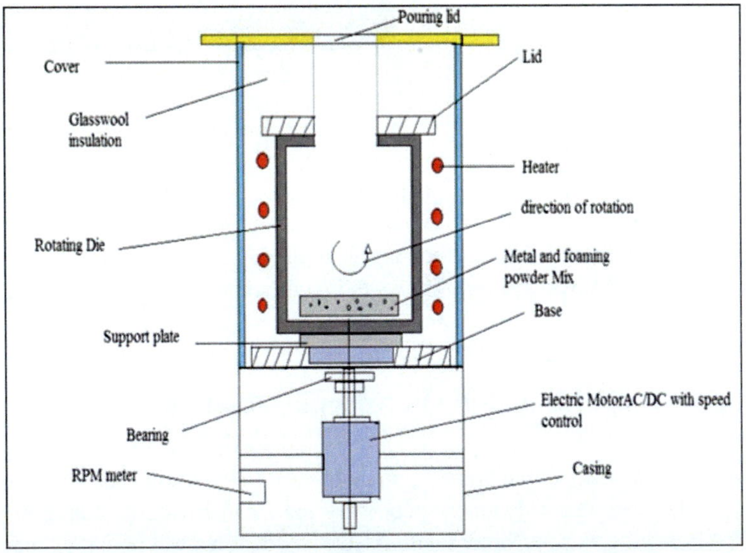

Fig. 1 Machine designed for small batch production of aluminum and its alloys

Table 1 Chemical analysis of aluminum alloy

Sr. no	Element name	Element %
1	Mn	0.16
2	Si	9.64
3	Cu	1.34
4	Fe	0.8
5	Zn	0.89
6	Ni	0.07
7	Pb	0.3
8	Al	Balance

Fig. 2 Al melting process

of aluminium alloy is shown in Table 1. Spectroscopic analysis of raw material is done and chemical analysis confirms to BS1490:1988 LM2 (AlSi10Cu2) EN AB-46100. This raw material is used for the foaming experiment by adding foaming agent $LiAlH_4$ in proportions like 5, 10, and 15% to find the effect of foaming agent on void formation. $LiAlH_4$ is in the powder form having an assay value of Min.96% is added in the alloy LM2. Figure 2 shows the Al melting process.

4 Sample Preparation for Microscopic Observation

Standard metallurgical sample preparation involves rough cutting and polishing, dry polishing with silicon carbide emery papers having mesh sizes 200, 400, 600, 800, 1000. Wet polishing was performed with Al_2O_3 powder on a wet polishing machine and etching with NH4OH.

From the experimentation work, following are the observations:

- A vertical axis centrifugal casting process is a very common casting process. Effect of centrifugal force during foam cell formation, expansion, and cell size are obtained.
- Microstructural analysis shows the morphology of the cell and the presence of voids. It is shown in Table 2.
- X-ray diffraction is a non-destructive method used to provide detailed information about the chemical composition, crystallographic structure, and physical properties of a material [14, 15]. Optical microscope observation reporst at 250X and 400X magnification. Microstructure is used to confirm the crystalline phases present in a material and reveal chemical composition information. The confirmation is done by using XRD data for aluminum and aluminum alloy available from standard resources as shown in Figs. 3, 4 and 5. The specifications of the machine are Goniometer Ultima IV Attachment Sample Rotation (with Z) Att. Scanning Mode 2Theta/Theta Scanning Type Continues Scanning X-Ray.40 kV/ 40 mA.

Table 2 Microstructure observation

Sr. no	Microstructure	Details	Remarks
1		LM2 alloy + 5% $LiAlH_4$ (Foaming Agent) at 400X magnification and NaOH etchant	35% voids
2		LM2 alloy + 10% $LiAlH_4$ (Foaming Agent) at 400X magnification and NaOH etchant	55% voids
3		LM2 alloy + 15% $LiAlH_4$ (Foaming Agent) at 400X magnification and NaOH etchant	65% voids

peak search

Sample	: AKashidS1	File	: AKashidS1.raw	Date	: October-07-22 15:57:54
Comment	: SolapurUniversityIC	Memo	: AKashid S1	Operator	: DrMakarandKulkarni
Method	: 2nd differential	Typical width	: 0.800 deg.	Min. height	: 100.00 cps

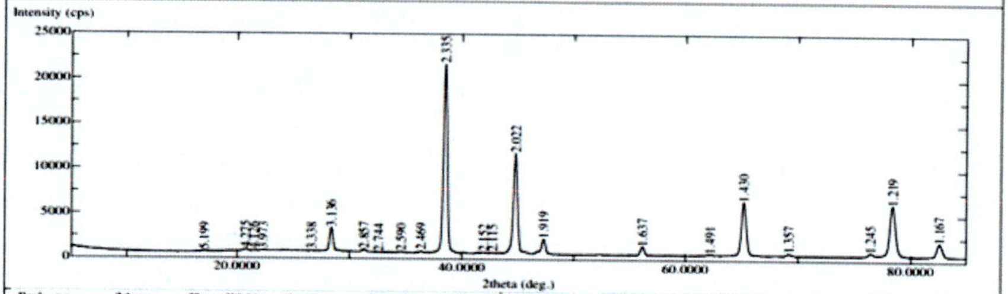

Peak no.	2theta	Flex Width	d-value	Intensity	I/Io
1	17.040	0.494	5.1992	811	4
2	20.760	0.518	4.2752	1007	5
3	21.520	0.424	4.1259	868	4
4	22.360	0.565	3.9727	896	5
5	26.680	0.424	3.3385	891	5
6	28.440	0.494	3.1357	3406	16
7	31.280	0.518	2.8572	1038	5
8	32.600	0.494	2.7445	758	4
9	34.600	0.494	2.5903	820	4
10	36.360	0.447	2.4688	876	5
11	38.520	0.518	2.3352	21664	100
12	41.940	0.424	2.1523	800	4
13	42.720	0.424	2.1148	786	4
14	44.780	0.518	2.0222	11837	55
15	47.320	0.494	1.9194	2299	11
16	56.140	0.518	1.6370	1556	8
17	62.200	0.541	1.4913	761	4
18	65.200	0.565	1.4297	6585	31
19	69.180	0.541	1.3568	874	5
20	76.420	0.541	1.2453	1024	5
21	78.380	0.612	1.2190	6270	29
22	82.580	0.612	1.1673	2119	10

October-07-2022 17:32:14 Page-1

Fig. 3 XRD pattern for LM2 alloy + 5% $LiAlH_4$

peak search

Sample	: AKashidS2	File	: AKashidS2.raw	Date	: October-07-22 16:40:26
Comment	: SolapurUniversityIC	Memo	: AKashid S2	Operator	: DrMakarandKulkarni
Method	: 2nd differential	Typical width	: 0.800 deg.	Min. height	: 100.00 cps

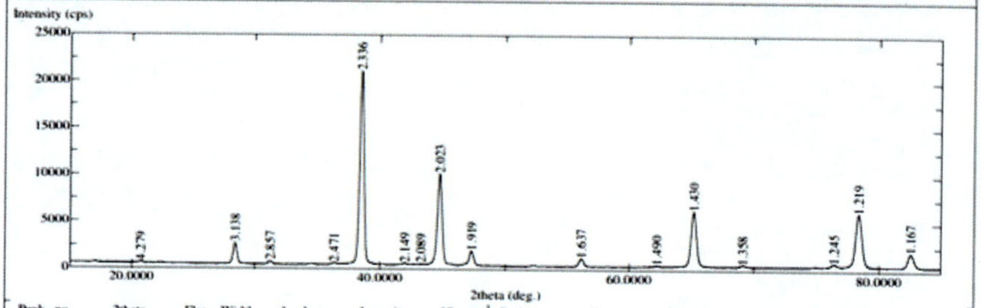

Peak no.	2theta	Flex Width	d-value	Intensity	I/Io
1	20.740	0.494	4.2792	681	4
2	28.420	0.494	3.1379	2676	13
3	31.280	0.494	2.8572	817	4
4	36.320	0.494	2.4715	685	4
5	38.500	0.518	2.3364	21106	100
6	42.000	0.471	2.1494	696	4
7	43.280	0.424	2.0888	616	3
8	44.760	0.518	2.0231	10200	49
9	47.320	0.518	1.9194	2000	10
10	56.140	0.541	1.6370	1289	7
11	62.240	0.447	1.4904	604	3
12	65.200	0.565	1.4297	6352	31
13	69.140	0.612	1.3575	690	4
14	76.420	0.565	1.2453	879	5
15	78.360	0.612	1.2193	6206	30
16	82.580	0.612	1.1673	2036	10

October-07-2022 17:33:29 Page-1

Fig. 4 XRD pattern for LM2 alloy + 10% $LiAlH_4$

peak search

Sample	: AKashidS3	File	: AKashidS3.raw	Date	: October-07-22 17:16:57	Operator	: DrMakarandKulkarni
Comment	: SolapurUniversityIC	Memo	: AKashid S3				
Method	: 2nd differential	Typical width	: 0.800 deg.	Min. height	: 100.00 cps		

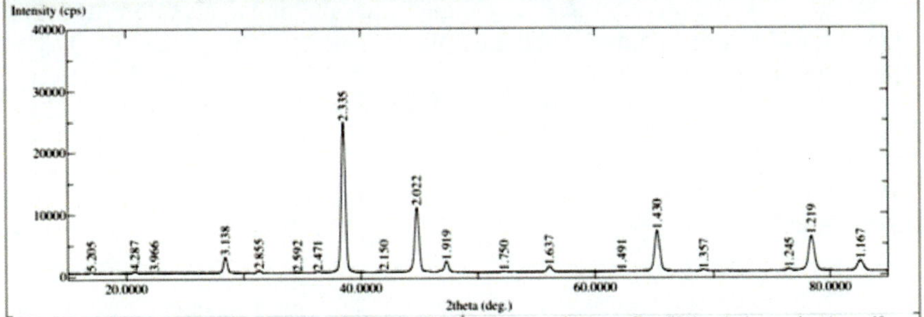

Peak no.	2theta	Flex Width	d-value	Intensity	I/Io
1	17.020	0.494	5.2052	760	4
2	20.700	0.494	4.2874	865	4
3	22.400	0.447	3.9657	718	3
4	28.420	0.494	3.1379	3166	13
5	31.300	0.518	2.8554	1024	5
6	34.580	0.659	2.5917	686	3
7	36.320	0.518	2.4715	778	4
8	38.520	0.518	2.3352	24931	100
9	41.980	0.447	2.1504	737	3
10	44.780	0.518	2.0222	11086	45
11	47.320	0.494	1.9194	2348	10
12	52.240	0.518	1.7496	647	3
13	56.140	0.541	1.6370	1460	6
14	62.220	0.588	1.4908	686	3
15	65.200	0.565	1.4297	7202	29
16	69.160	0.729	1.3572	777	4
17	76.420	0.541	1.2453	955	4
18	78.360	0.612	1.2193	6249	26
19	82.600	0.612	1.1671	2298	10

October-07-2022 17:42:51 Page-1

Fig. 5 XRD pattern for LM2 alloy + 15% $LiAlH_4$

5 Conclusions

- A centrifugal casting method can be combined with the powder metallurgy precursor method of foam manufacturing. Powder metallurgy precursor method is a commonly used foam manufacturing process. The centrifugal force offered during centrifugal casting evenly distributes the melt and prevents floating of lighter phases upward. This combination leads to new technology in foam manufacturing.
- A low-cost $LiAlH_4$ foaming agent can replace TiH2 foaming agent.
- This method is suitable for batch production of cylindrical-shaped products in less time.
- The overall cost of production is less compared to the bubbling method, which is a commonly used method. Precursor of the desired composition can be produced on a mass scale in less time. These can be stored in plastic bags for long shelf life. Melting is done in a rotating mold electric muffle furnace, which is energy efficient and of low energy cost.

- Performance: Material composition is exactly obtained as powder metallurgy precursor. Melting is uniform and homogenous. High foaming rate is obtained at 60–90 rpm.
- Accuracy: The chemical composition, dimensions, melting temperature, and rotational speed can be monitored and controlled precisely using sensors for desired foam density.

References

1. Neu TR, Pfretzschner B, Garcia-Moreno F, Banhart J (2017) Influence of the heating rate on the foaming behaviour of various aluminium alloys. Metals 7(2017):323. https://doi.org/10.3390/met7090323
2. Banhart J (2001) Manufacture, characterization and application of cellular metals and metal foams. Prog Mater Sci 46:559–632
3. Liu X, Li Y, Chen X, Yuan L, Fan X (2010) Foam stability in gas injection foaming process. J Mater Sci 45:6481–6493. https://doi.org/10.1007/s10853-010-4736-5
4. Sánchez-Martínez A, Cruz A, González-Nava M, Suárez MA (2016) Main process parameters for manufacturing open-cell Zn-22Al-2Cu foams by the centrifugal infiltration route and mechanical properties. Mater Design 108:494–500
5. Deng-wei H, Juan Y, Xiang-yang Z, Hui W, Zhang T (2012) Preparation of open-celled aluminum foams by counter-gravity infiltration casting, Trans Nonferrous Met Soc China 22:85–89
6. Goodall R, Despois OJF, Marmottant A, Salvo L, Mortensen A (2006) The effect of preform processing on replicated aluminium foam structure and mechanical properties. Scripta Mater 54:2069–2073
7. Rajak DK, Kumaraswamidhas LA (2017) Technical overview of aluminum alloy foam. Rev Adv Mater Sci 48
8. Banhart J (2006) Metal foams: production and stability. Adv Eng Mater 8(9):781–794
9. Banhart J, Baumeister J (1998) Production methods for metallic foams. Mat Res Soc Symp Proc 521:121–133
10. Matijasevic B, Banhart J(2006) Improvement of aluminum foam technology by tailoring of blowing Agent. Scripta Mater 54(4):503–508
11. Wannasin J, Flemings MC (2005) Fabrication of metal matrix composites by a high-pressure centrifugal infiltration process. J Mater Process Technol 169:143–149
12. Baltatescu O, Axinte M, Barbu G, Manole V (2015) New approach for porous material obtaining using centrifugal casting. Modern technologies in industrial engineering (MedTech 2015) 10P Conf. Series: Material science and engineering 95
13. Baumgartner F, Durate I, Banhart J (2000) Industrialization of powder compact foaming process. Adv Eng Mater 2(4):168–174
14. Gibson LJ (2000) Metallic foams: structure, properties and applications annual review of material science. In: Aref H, Phillips JW (eds) Mechanics for a new Mellennium. Springer, Dordrecht, pp 57–74. https://doi.org/10.1007/0-306-46956-1_4
15. Banhart J, Baumeister J, Weber M (1995) Powder metallurgical technology for the production of metallic foams. In: Proceedings of the European conference on advanced PM materials (PM'95) Birmingham (UK), 23.-25.10.1995 (European Powder Metallurgy Association, Shrewsbury), pp. 201–208

Chapter 12
Synthesis and Characterization of Nano-CuO Complexed HPMC:PVA Polymer Blend Electrolytes

N. Sandhya Rani, Sunil Kumar, B. Vinod, C. S. Dileep, C. Manasa, and N. Srikantamurthy

Abstract Polymerelectrolyte films of polyvinylalcohol (PVA)/hydroxypropyl methylcellulose (HPMC) complexed with copper oxide (CuO) nanoparticles (1%–4% wt.%) were synthesized by adopting casting solution technique. X-ray diffraction results showed a reduced percentage of crystallinity with the increase in CuO concentration. Scanning electron microscopy (SEM) showed the different morphology of all the samples. Differential scanning calorimetry (DSC) analysis shows the increase in glass transition as well as in melting temperature for all samples. This is because the presence of CuO nanoparticles in the host polymer blend matrix makes them more thermally stable. The tensile strength and modulus nano-CuO complexed samples show improvement when compared with HPMC:PVA blends. The AC conductivity also increases with frequency and CuO concentration. All the experimental studies showed that 3 wt% nano-CuO complexed HPMC:PVA polymer electrolyte samples are more thermally stable and exhibit high conductivity at room temperature.

Keywords Nanocomposite polymer electrolytes · X-ray diffraction · DSC

N. Sandhya Rani (✉)
Department of Applied Physics, VVCE, Mysore, India
e-mail: sandhya.phy@vvce.ac.in

S. Kumar
Department of Physics, S.S Arts College and T.P Science Institute, Sankeshwar, Belagavi 591313, India

B. Vinod
Department of Mechanical Engineering, VVCE, Mysore, India
e-mail: vinod@vvce.ac.in

N. Sandhya Rani · C. S. Dileep · C. Manasa · N. Srikantamurthy
Department of Applied Chemistry, VVCE, Mysore, India
e-mail: srijmn@vvce.ac.in

V. Bindhu et al. (eds.), *Proceedings of Sixth International Conference on Inventive Material Science Applications*, Advances in Sustainability Science and Technology,
https://doi.org/10.1007/978-981-99-4189-6_12

1 Introduction

One of the key constituents of the electrochemical cell devices is polymer electrolyte [1–4]. It is made by dissolving metal salt in a polymer host [5–7]. Here, an environmentally friendly polymer mixture of HPMC and PVA is employed as the host polymer matrix to boost electrical conductivity and give extra sites for ion hopping and ion transfer as hydrogen bonds are formed between the polymer chains [8–12]. To improve the thermal, structural, mechanical, and electrical conductivity qualities of the synthesized polymer electrolyte films, CuO nanoparticles are added to the polymer mixture [13–18].

2 Experimental Work

2.1 Materials and Sample Preparation Method

The HPMC and PVA solutions (5:5 wt%) were made separately, combined, and stirred continuously. This blend solution was gradually added with CuO nanoparticles (1–4 wt%), which were then continuously swirled for two hours at 380 rpm using a magnetic stirrer. After being evaporated in a vacuum chamber for a period of ten minutes, the liquid was placed into flat plates to cool down at the ambient temperature for four to five days. Film samples were coded as HPMC:PVA blend (S1), HPMC:PVA—1% CuO (S2), HPMC:PVA—2% CuO (S3), HPMC:PVA—3% CuO (S4), HPMC:PVA—4% CuO (S4).

2.2 Instrumentation

X-ray diffractometer with CuK radiation (1.5406), a Rigaku Miniflex-II was used to obtain the XRD patterns on a scale of $10° - 80°$. The SEM studies were done by using an electron microscope (JEOL 840) with a resolution of 20 kV. DSC-Q20, Universal V4.5 A was used to create the DSC thermograms. The tensile test was done on the test specimens prepared as per ASTM D-638 (165*19*6) type 1 standard (Fig. 4). The AC conductivity was measured using an LCR meter ZM2376 at operating frequencies ranging from 50 Hz to 1 MHz. Using Eq. 1, the electrical conductivity is determined.

$$\sigma = \left(\frac{1}{R_b}\right)\left(\frac{t}{A}\right) \tag{1}$$

where 'A' is the area mm^2, 't' is the thickness in mm, and 'R_b' is the resistance in MΩ.

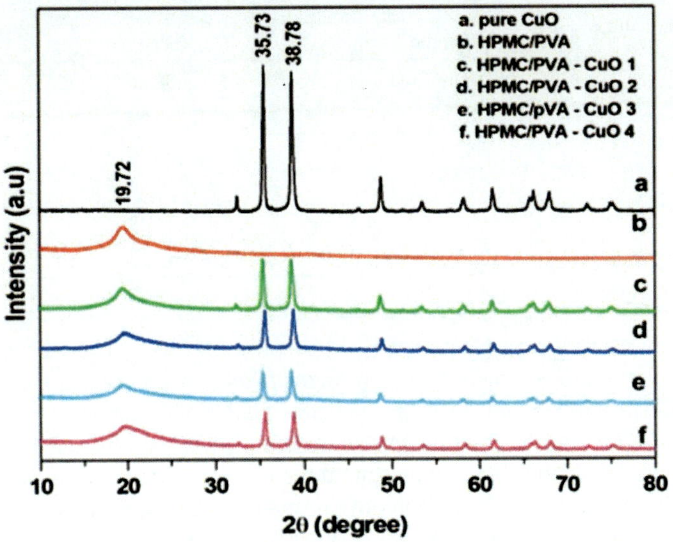

Fig. 1 XRD pattern of all samples

3 Results and Discussion

3.1 X-Ray Diffraction Studies

Figure 1 displays the XRD graphs of all samples. According to well-reported literature, the electrical conductivity of the polymer electrolyte sample is higher in the amorphous area [19–23]. Peaks for pure Cuo are more powerful and distinct at about 34.72°, 38.82°and 47.70°. While the HPMC:PVA blends show a visible broad peak at 19.69°. The sharp peak of pure CuO and the broad peak of polymer blends reduce in all other polymer electrolyte samples; hence, the percentage of crystallinity also reduces (Table 1). Percentage of crystallinity is calculated using Eq. 2.

$$\text{Crystallinity}(\%) = \frac{\text{crystalline peaks area}}{\text{crystalline peaks area} + \text{amorphous haloes area}} \text{X100} \tag{2}$$

3.2 Morphological Studies

Figure 2 displays the film morphologies of the reported polymer electrolyte sample. Due to the absence of discernible crystalline regions, pure HPMC and pure PVA

Table 1 Crystalline parameters

Sample	X_C(%)
Pure CuO	58.66
S1	18.52
S2	56.60
S3	49.20
S4	31.43
S5	43.22

display semi-crystallanity [21, 22, 24]. SEM picture (Fig. 2c) reveals an evenly dispersed microporous structure. This demonstrates that one of the key elements affecting the polymer blend performance is the similarity of the interfaces between the segments [25]. Figure 2d–g shows that overall the CuO nanoparticles are widely disseminated throughout the polymer mixtures and interacted with them [26, 27]. This clearly demonstrates the viability of this approach to create a uniform layer of well-distributed nanoparticles [28–30].

3.3 DSC Studies

The thermal characteristics of the reported samples are displayed in Fig. 3. Table 2 provides an overview of the thermal properties obtained from these thermograms. Nano-CuO complexing with HPMC:PVA blends affects the melting temperature as well as the crystallization behavior of the polymer electrolytes. From XRD, a drop in the percentage of crystallinity was also observed for all the CuO complexed samples. It is clear from the DSC plots that the glass transition temperature T_g and melting temperature T_m increases slightly, whereas crystallization temperature T_c decreases slightly [31, 32]. All the values showed that the 3wt% Cuo complexed polymer blend had got high thermal stability.

3.4 Tensile Properties

Figure 4 shows the test specimen used for tensile test according to ASTM standars. Figure 5 displays the stress–strain curves of the reported samples. Table 3 shows the test results. The tensile strength of PVA:HPMC blend is 3.477 Mpa. The highest tensile strength of 13.533 Mpa is obtained for 3% CuO complexed sample. Therefore, its strength is almost 3.89 times greater than the pure polymer blend. Similarly, the modulus of the 3% CuO complexed sample is 0.162 Gpa, which is 65% greater than the polymer blends. It is evident that when the nano-CuO concentration rises, both the tensile strength and the modulus also rise and reaching to their maximum

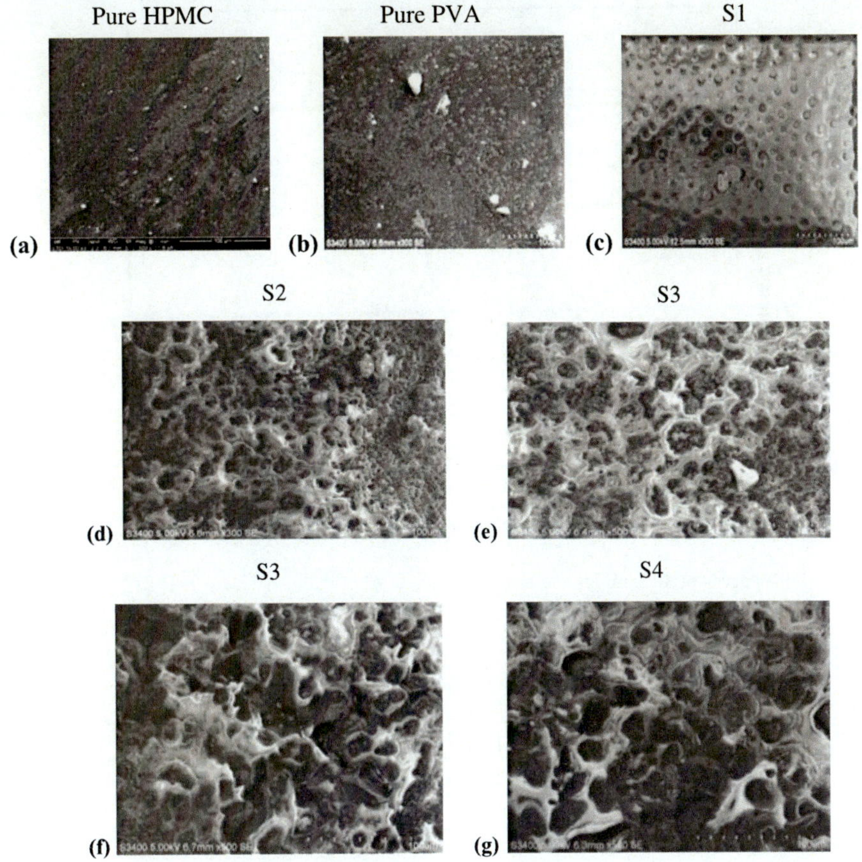

Fig. 2 SEM photographs of the samples

values at 3% CuO complexed HPMC:PVA blends. Whereas a higher concentration, i.e., for HPMC:PVA—CuO 4%, both tensile strength and modulus start to reduce. This exhibits that the nanocomposite sample of HPMC:PVA—CuO 3% has got more strength than other ratios of nano-CuO complexed samples.

3.5 *Eletrical Conductivity Studies*

Figure 6 clearly displays enhancement in the conductivity up to 3 wt% CuO and then decreases. It can also be seen that the HPMC:PVA—3 wt% CuO sample exhibits the highest ionic conductivity of 1.23×10^{-4} S/cm at room temperature. According to

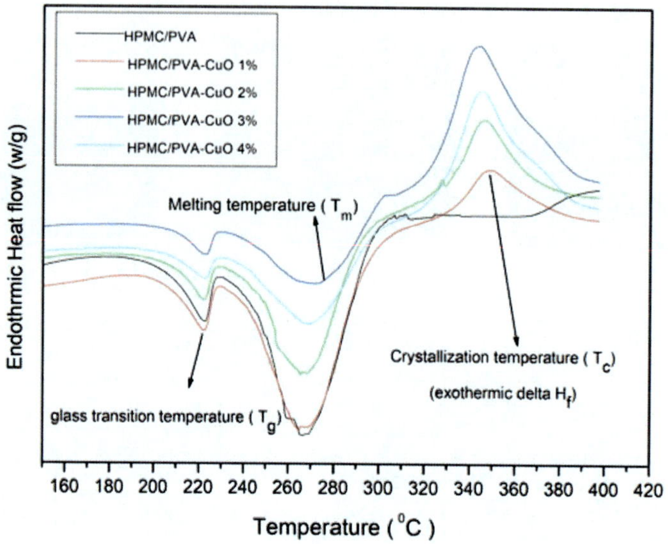

Fig. 3 DSC curves of all samples

Table 2 DSC parameters

Sample	T_g (°C)	T_m (°C)	$\Delta H_m (J/g)$	T_C (°C)
S1	220.11	264.17	328	–
S2	221.1	269.89	265	349.882
S3	223.18	268.78	222	347.20
S4	223.48	272.51	274	344.06
S5	222.43	269.56	270	344.83

Table 3 Tensile properties of the samples

Sample	Tensile strength (Mpa)	Modulus (Gpa)	Peak load (kN)	Stiffness kN/mm
S1	3.477	0.097	0.105	0.02
S2	3.659	0.102	0.107	0.02
S3	10.259	0.123	0.494	0.039
S4	13.533	0.162	1.021	0.082
S5	10.41	0.125	0.795	0.064

this, an increase in nano CuO dissociation is what is responsible for the increase in ionic conductivity as it produces more mobile ions. On the other hand, at larger salt concentrations, the dissociated ions start to recombine to create neutral salt, which reduces electrical conductivity [33, 34].

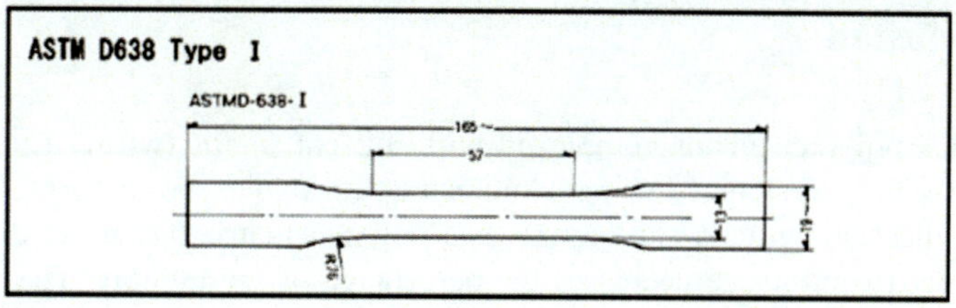

Fig. 4 Test specimen for tensile test (ASTM D638)

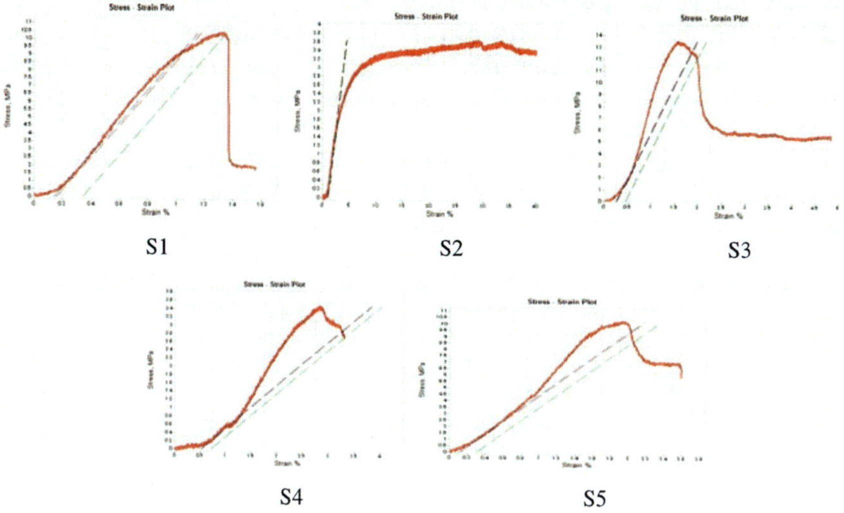

Fig. 5 Stress–strain curves of the samples

Fig. 6 Ionic conductivity behaviour of HPMC:PVA—CuO films

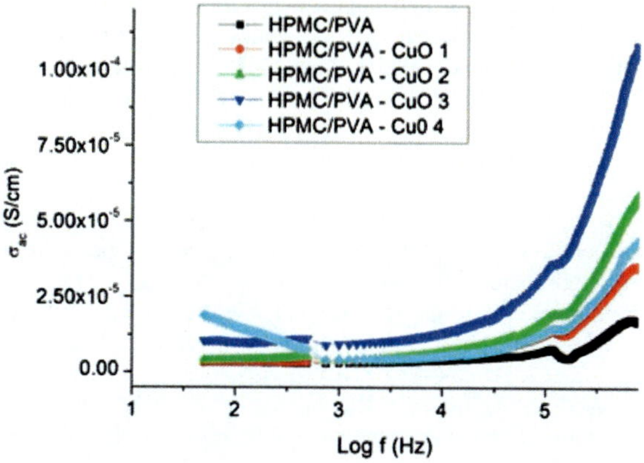

4 Conclusion

HPMC:PVA polymer blends complexed with diffrentt ratios (wt%) of nano-CuO were successfully syntehized using solution cast technique. From XRD studies, it is evident that adding nano-CuO to the host polymer matrix has an impact on the polymer matrix and hence decreases the percentage of crystallinity. The samples' rise in amorphous domain leads to an increase in the polymer electrolytes' electrical conductivity. SEM image of all the samples showed a microporous structure that is equally distributed. DSC studies show that the glass transition temperature, as well as melting temperature improved after nano-CuO complexing in the polymer mix, making them more thermally stable. Tensile and modulus studies showed good improvement in all the polymer electrolyte samples. The electrical conductivity also increases with increasing nano-CuO concentration. From all the studies, it was observed that the sample HPMC:PVA—CuO 3% shows more thermal stability, less crystallinity, and high conductivity of 1.28 1.28×10^{-4} S/cm at room temperature; hence it can become useful in electrochemical device applications.

Acknowledgements The authors are thankful to the LCR metre and computerized universal testing machine facilities provided by the Engineering Physics and Mechanical Engineering department of VVCE in Mysore for the investigation of mechanical characteristics.

References

1. Parameswaran V, Nallamuthu N, Devendran P, Nagarajan ER, Manikandan A (2017) Electrical conductivity studies on ammonium bromide incorporated with zwitterionic polymer blend electrolyte for battery application. J Phys B 515:89–98
2. Senthil RA, Theerthagiri J, Madhavana J, Arof AKM (2016) High performance dye-sensitized solar cell based on 2-mercaptobenzimidazole doped poly(vinylidinefluorideco-hexafluoropropylene) based polymer electrolyte. J Macromol Sci A 53:245–251
3. Senthil RA, Theerthagiri J, Madhavan J (2015) Hematite Fe2O3 nanoparticles incorporated polyvinyl alcohol based polymer electrolytes for dye-sensitized solar cells. J Mater Sci Forum 832:72–83
4. Hafiza MN, Bashirah ANA, Bakar NY, Isa MIN (2014) Electrical properties of carboxyl methyl-cellulose/chitosan dual-blend green polymer doped with ammonium bromide. J Int J Polym Anal Charact 19:51–58
5. Shukur MF, Ithnin R, Illias HA, Kadir MFZ (2013) Proton conducting polymer electrolyte based on plasticized chitosan-PEO blend and application in electrochemical devices. J Opt Mater 35(18):34–41
6. Abdullah OG, Hanna RR, Salman YAK (2017) Structural, optical, and electrical characterization of chitosan: methylcellulose polymer blends based film. J Mater Sci Mater Electron 28(102):83–94
7. Acosta JL, Morales E (1996) Structural morphological and electrical characterization of polymer electrolytes based on PEO/PPO blends. J Solid State Ionics 85:85–90
8. Yang CC, Wu GM (2009) Study of microporous PVA/PVC composite polymer membrane and it application to MnO2 capacitors. J Mater Chem Phys 114:948–955

9. Sandhya Rani N, Sannappa J, Demappa T, Mahadevaiah (2014) Structural, thermal, and electrical studies of sodium iodide (NaI)-doped hydroxypropyl methylcellulose (HPMC) polymer electrolyte films. J Ionics 20(2):201–207
10. N. Sandhya Rani, J. Sannappa, T. Demappa, Mahadevaiah (2015) Effects of CdCl2 concentration on the structural, thermal and ionic conductivity properties of HPMC polymer electrolyte films. J Ionics 21:133–140
11. Nouh SA, Abdel-Kader MH, Mohamed MB (2017) Structural and optical modifications in polyvinyl alcohol due to Cr2O3 nanoparticles additives concentration, and gamma irradiation. J Adv Polym Technol 3:1–5
12. Mohammed G, Morsi WM (2018) Spectroscopic, thermal and electrical properties of MgO/PVP/PVA nanocomposites. J Phys Chem Solids 115:238–247
13. Tamgadge YS, Talwatkar SS, Sunatkari AL, Pahurkar VG, Muley GG (2015) Studies on nonlocal optical nonlinearity of Sr–CuO–polyvinyl alcohol nanocomposite thin films. J Thin Solid Films 595:48–55
14. Sivasubramanian R, Biji P (2016) Preparation of copper (I) oxide nanohexagon decorated reduced graphene oxide nanocomposite and its application in electrochemical sensing of dopamine. J Mater Sci Eng 2(10):10–18
15. Souza SD, Hidembergue O, Sanches EA (2018) Polyaniline-CuO hybrid nanocomposite with enhanced electrical conductivity. J Mol Struct 1153:20–27
16. Jundale DM, Navale ST, Khuspe GD, Dalavi DS (2013) Polyaniline–CuO hybrid nanocomposites: synthesis, structural, morphological, optical and electrical transport studies. J Mater Sci Mater Electron 24:3526–3535
17. Alghunaim NS (2019) Effect of CuO nanofiller on the spectroscopic properties, dielectric permittivity and dielectric modulus of CMC/PVP nanocomposites. J Mater Resea Technol 8(4):3596–3602
18. Ali Alhazime A (2020) Efect of nano CuO doping on structural, thermal and optical properties of PVA/PEG blend. J Inorg Organ Polym Mater 157:7–8
19. Ahmed HT, Abdullah OG (2020) Structural and ionic conductivity characterization of PEO:MC-NH4I proton-conducting polymer blend electrolytes based films. J Resul Phy 16:102861
20. Abdullah OG, Aziz SB, Rasheed MA (2016) Structural and optical characterization of PVA:KMnO4 based solid polymer electrolyte. J Resul Phys 6(110):3–8
21. Ahmed HT, Abdullah OG (2019) Preparation and composition optimization of PEO:MC polymer blend films to enhance electrical conductivity. J Polym 11:853
22. Singh NL, Shah S, Qureshi A, Tripathi A, Singh F, Dkavasthi. J. Bull. Mater. Sci. 34, 81–88 (2011)
23. Mohamada AH, Abdullahb OG, Saeed SR (2020) Effect of very fine nanoparticle and temperature on the electric and dielectric properties of MC-PbS polymer nanocomposite films. J Resul Phys 16(102898):1–9
24. Bianchia SE, Angelia VW, de Souzaa KCB (2011) Evaluation of the solubility of the HPMC/PVA blends in biological fluids in vitro. J Mater Resea 14(2):166–171
25. Jundale D, Pawar S, Chougule M, Godse P, Patil S, Raut B (2011) Nanocrystalline CuO thin films for H2S monitoring microstructural and optoelectronic characterization. J Sens Technol 36–46
26. Jundale DM, Joshi PB, Sen S, Patil VB (2012) Nanocrystalline CuO thin films: synthesis, microstructural and optoelectronic properties. J Mater Sci Mater Electron 23:1492–1499
27. Lee KH, Kim HY, La YM, Lee DR, Sung NH (2002) Influence of a mixing solvent with tetrahydrofuran and N, N-dimethylformamide on electrospun poly(vinyl chloride) nonwoven mats. J Polym Sci Polym Phys 40:2259–2268
28. Han YG, Kusunose T, Sekino T (2009) One-step reverse micelle polymerization of organic dispersible polyaniline nanoparticles. J Synth Mater 159:123–131
29. Jundale DM, Navale ST, Khuspe GD, Dalavi DS, Patil PS, Patil VB (2013) Polyaniline–CuO hybrid nanocomposites: synthesis, structural, morphological, optical and electrical transport studies. J Mater Sci Mater Electron 24:3526–3535

30. Sahoo S, Hajra S, Mohantha K, Choudhary RNP (2018) Processing dielectric and impedance spectroscopy of lead free. J Alloy Comp 766:25–32
31. Falqi FH, Bin-Dahman OA (2018) Preparation of miscible PVA/PEG blends and effect of graphene concentration on thermal, crystallization, morphological, and mechanical properties of PVA/PEG (10wt%) blend. J Hindawi Int J Polym Sci 1–10
32. Ali ZI, Eisa WH (2014) Characterization of electron beam ırradiated poly vinyl alcohol/poly ethylene glycol blends. J Sci Res 6:29–4
33. Alakanandana A (2017) Structural and electrical characterization of PVA doped malonic and succinic acid polymer electrolytes. J Int J Sci Tech Eng 3(7):30–36
34. Schantz S, Torell LM (1993) Evidence of dissolved ions and ion pairs in dilute poly (propylene oxide)-salt solutions. J Solid State Ionics 60:47–53

Chapter 13
Study on Effect of Pista Shell as Filler Material on Mechanical Performance of Banana Fiber Reinforced Polymer Composite

B. Vinod, L. J. Sudev, B. B. Ganesha, K. N. Arunkumar, and C. S. Thammegowda

Abstract Recently, more research efforts have been initiated to improve the performance of composite materials due to their ever-increasing demand. A range of pre- and post-processing processes may be used to modify the biological, chemical, physical, and mechanical properties of any material. As a result, the primary goal of this study is to investigate the effect of powdered pistachio shells as a filler on the mechanical performance of banana fiber-reinforced polymer composites. The 10 wt.% of filler material, 30 wt.% of fiber and 60 wt.% of epoxy resin were used to prepare composite laminates using compression mould technique for rigid compaction of fiber and filler material. From experimental investigation, it was observed that the use of filler material improved the hardness and stiffness of the composite. The tensile stress was reduced to 22.97 MPa from 25.34 MPa and flexural strength was reduced to 36.76 MPa from 43.66 Mpa, respectively. The Young's modulus was increased to 4.96 GPa from 3.23 GPa and flexural modulus was increased to 7.38 GPa from 6.21 Gpa, respectively.

Keywords Banana · Filler · Epoxy Resin · Hardness · Stiffness · Strength

B. Vinod (✉) · L. J. Sudev · B. B. Ganesha · K. N. Arunkumar · C. S. Thammegowda
Department of Mechanical Engineering, VVCE, Mysuru, India
e-mail: vinod@vvce.ac.in

L. J. Sudev
e-mail: sudevlj@vvce.ac.in

B. B. Ganesha
e-mail: bbganesh.bbg@vvce.ac.in

K. N. Arunkumar
e-mail: arunkn.10@vvce.ac.in

C. S. Thammegowda
e-mail: thammegowda.cs@vvce.ac.in

V. Bindhu et al. (eds.), *Proceedings of Sixth International Conference on Inventive Material Science Applications*, Advances in Sustainability Science and Technology,
https://doi.org/10.1007/978-981-99-4189-6_13

1 Introduction

When it comes to Fiber Reinforced Polymers (FRP) composites, there may be a possibility of an interstitial space between the fibers and the matrix material. This interstitial space may act as a void or inner crack, thereby reducing the performance ability of composite material. Therefore, it is essential to reduce the space between the fibers and matrix material through various processes during manufacturing. Applying the compressive force during manufacturing is one of the effective methods to reduce the inner voids in composite. At the same time, application of compressive force may generate internal stress in the composite material, which further reduces the load-bearing ability of the composite. Therefore, one of the best methods to reduce the space between matrix and fiber is the utilization of filler materials. These materials can either be available naturally or it could be manmade. Most commonly used natural fillers are from agricultural waste like pistachio shell, rice husk, wheat husk, ash and coconut coir, etc. There are certain inorganic materials that are used as filler materials like calcium carbonate, calcium sulfate, alumina trihydrate and kaolin (hydrous aluminum silicate) etc. [1].

Iskender Ozsoy et al. [2] made an investigation on the effect of micro- and nanofillers such as Al_2O_3, TiO_2, clay and fly ash on epoxy composite for 10–30wt.% by weight ratios. From experimental investigation, it was examined that the tensile and flexural strength of the composite decreases with increasing micro- and nanofiller content ratio. However, the modulus of elasticity during tensile and bending tests improves with the increase of the content ratio of micro- and nanofillers. Hardness of the composite increased with the addition of filler material and thus making the composite brittle. Zaman et al. [3] analysed the micro- and nano-ZnO-stuffed polypropylene composites for different weight ratios ranging between 2 and 8wt.%. The study inferred that nano-filled composites exhibits better strength than micro-filled composites for the same filler ratios. The 5 wt.% filler content was found to be an optimum ratio for which the dispersion of particles was better. A stronger interfacial bonding between the matrix and fillers was observed with the help of morphology images. Gao et al. [4] examined an effect of nano-sized $CaCO_3$ as filler on the mechanical strength of polystyrene composites. The stiffness starts to increase with increasing the filler ratio up to a certain extent, beyond which the filler starts to agglomerate, thereby reducing the adhering ability between matrix and fiber material. Thus, the strength of the composite starts to decrease. Agubra et al. [5] studied the impacts of nano-clay distribution procedures on the mechanical performance glass fiber epoxy composites. Magnetic dispersion, planetary centrifugal mixing and ultrasonication are some of the commonly used dispersion methods to attain moderate dispersion of nanoparticles in polymer matrix. In this research, the influence of the three roll milling method for nano-clay dispersion into epoxy matrix was analyzed. The paper concluded that significant clay dispersion in the epoxy matrix was accomplished by merging the magnetic stirring and tinky mixing followed by three roll milling method. Yasmin et al. [6] investigated the structural, mechanical, viscoelastic and thermal behavior of graphite-filled epoxy composites of filler weight ratios between 2.5 and 5 wt.%. The tensile strength, glass transition temperature and modulus increased with increasing graphite platelets up to 5 wt.% filler ratios. However, above which, the agglomeration of fillers occurred. The coefficient of thermal expansion was reduced with the increase of graphite platelets and also there was an improvement in the thermal stability of the composite. Sayer [7] studied the effect of ceramic fillers like SiC, Al_2O_3, and B4C on the elastic- and load-carrying ability of e-glass fiber reinforced epoxy resin. The filler weight ratios were varied from 5 to 15% wt. The article concluded that the 10% Wt. attained as the optimal value the composite exhibiting maximum load-bearing ability and elastic moduli. Apart from the weight percentage, it was also reflected that the particle size also plays a major role to improve the strength of the composite. Smaller the particle size greater will the elastic moduli and load-carrying ability of the composite. Asi [8] studied the mechanical properties of glass fiber-reinforced epoxy composites filled with various proportions of Al_2O_3. Experimental results showed that the tensile and shear strengths decreased with the increasing addition of Al_2O_3 particles. However, the flexural strength of the composite was increased by 33% and flexural modulus by 78% with an increase in Al_2O_3 particles up to 10% wt. beyond which it starts to decrease. Yang et al. [9] analyzed the effect of lignocellulosic material (rice husk flour) as fillers on the mechanical properties of thermoplastics composite (polypropylene). The composite was prepared for different filler loading of 10, 20, 30 and 40% weight ratios. The paper concluded that interfacial area improves with an increase in filler ratios and thereby increasing the stiffness, hardness, and brittleness of the composite.

The comprehensive literature survey on the pertaining topic has revealed enormous information related to filler materials and helps to identify the gap that exists and to move in the right path. Therefore, in the current work, the effect of the powdered pistachio shell as filler material for weight ratios ranging from 0 to 10% wt. on the mechanical properties of banana fiber reinforced epoxy composites was studied. The tensile, flexural and hardness tests were performed, and the results were analyzed.

2 Materials and Methodology

The details on materials, processing and the experimental procedures followed in the present work are discussed below.

2.1 Banana Fiber

Banana fibers are well-known natural fibers to mankind from past decades. India is the world's largest producer of banana fruit, followed by China, Brazil, the Philippines and Ecuador. The excellent blending ability of banana fibers have extended its value-added applications in various fields [10]. It is a lignocellulosic fiber with very good

Table 1 The chemical compositions of banana fiber [10]

Chemical composition	Range in percentage
Cellulose	63–64
Hemicellulose	16–19
Lignin	5
Moisture content (%)	10–11
Density (g/cc)	1.35
Microfibrillar angle (deg)	11

strength obtained from the stem of the banana plant (Musa sepientum). The cellulose in the fiber acts as a cement or binder that gives the fiber its strength. Together with cellulose, the lignin content of the cell wall increases its stiffness and hemicellulose strengthens the cell wall due to its interaction with lignin and cellulose [11]. Table 1 shows the chemical compositions of banana fiber. A small variation in the assessments of these properties may be noted due to the difference in the source from where the fibers are available, the process of fiber extraction, age of the plant and climatic conditions, etc.

2.2 *Pistachio Shell (Filler)*

Pistachio is a small tree originating from Central Asia which belongs to the cashew family. After extraction of nuts, pistachio shells are commonly used for mulching, potted plant fillers, pest deterrent, fire kindling and compost, etc. Pistachio shells possess high stiffness, strength and hardness. Even though it is least considered as filler/reinforcing material in the case of fiber reinforced polymer composite materials [12].

2.3 *Lapox L12 Epoxy Resin*

Lapox L12 is an unmodified epoxy resin of medium viscosity, also known as poly-epoxides based on Diglycidyl Ethers of Bisphenol A (DGEBA). Diethylene triamine (DETA) was used to cure lapox L12, which is a light-yellow aliphatic polyamine. The reaction of the poly-epoxide with the multifunctional curing agent forms a thermoset polymer with excellent mechanical properties, adhering ability, dimensional stability, thermal insulator, fire retardance, low cure shrinkage, resistance to corrosion and water ingression, etc. [13]. Lapox L12 is commonly used for marine coating, flooring, aerospace, automobile and packaging industries, etc.

2.4 Methodology

The raw banana fibers were extracted from the trunk of the banana plant by retting process. The obtained raw fibers contain impurities and fatties at its surface. Therefore, in order to remove these impurities, a pretreatment operation of fibers such as an alkali treatment and drying the fibers using a hot air oven has been performed [14]. In the case of alkaline chemical treatment, the fibers were immersed in 5% NaOH solution for 1 h and then completely dried in a hot air oven at 60 °C to remove internal moisture. The finished fiber is spun into a mat and laminated by hand with a constant volume fraction of 30% fiber and 60% resin. The composite laminates were prepared by a 2% increase in the weight fraction of filler material up to 10%.

Test specimens were prepared according to American Society for Testing and Materials (ASTM) standards D638 and D790 for tensile and flexural tests, respectively. Similarly, hardness and impact testing were carried out according to ASTM D2240 and D256, respectively. High-speed waterjet cutting is used to cut samples into desired shapes [15]. Testing was performed using a computerized universal testing machine (UTM) with a constant load of 1kN and with a rate of elongation of 0.3 mm/min until the specimen broke. The detailed dimension of ASTM D638 and ASTM D790 are shown in Figs. 1 and 2, respectively.

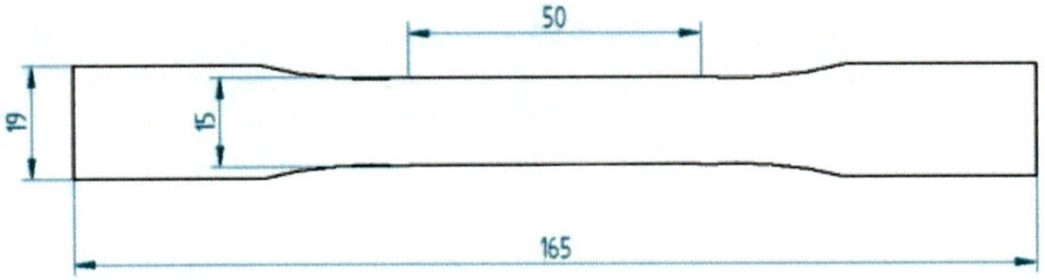

Fig. 1 Dimensions of tensile test specimen as per ASTM D638 (in mm)

Fig. 2 Dimensions of flexural test specimen as per ASTM D790 (in mm)

3 Results and Discussion

The specimen with different combinations of fiber and filler materials were subjected to tensile, flexural, impact and hardness test. The specimen with 70% wt. of Resin (R), 30% wt. of fiber (F) and 0% wt. of filler material was represented as sample A and similarly as B, C, D, E and F for the combinations as mentioned in Table 1, respectively. From experimental investigation, it is very much clear that the mechanical behavior of composite material varies by varying the amount of filler material. Table 2 shows the tensile, flexural, impact and hardness test results of the specimen with and without filler material. The values given in Table 2 are the average of 5 tests and the tests were conducted until failure of the sample occurred. Figure 3 shows the variation of tensile strength and tensile modulus of the specimens.

The test results clearly show that the percentage of filler material greatly influences the tensile strength and Young's modulus of the composite material. The maximum tensile strength of 41.76 MPa was achieved for sample A (i.e.,70% of resin, 30% of fiber and 0% of filler material). The tensile stress of the composite laminate starts to decrease gradually with the increase in the percentage of filler material, as shown in Fig. 3a. The tensile loading ability of the composite was reduced by 3.11%, with a

Table 2 Mechanical properties of composites with and without filler material

Sample	Combinations	Volume %	Tensile strength (MPa)	Young's modulus (GPa)	Flexural strength (MPa)	Flexural modulus (GPa)	Impact strength (KJ/m^2)	Hardness
A	Resin	70	41.76	2.98	84.56	3.69	6.41	114
	Fiber	30						
	Filler	0						
B	Resin	70	40.46	3.01	78.62	3.84	4.11	116
	Fiber	28						
	Filler	2						
C	Resin	70	38.45	3.15	75.02	3.97	3.52	119
	Fiber	26						
	Filler	4						
D	Resin	70	36.08	3.65	69.48	4.65	3.05	126
	Fiber	24						
	Filler	6						
E	Resin	70	32.53	3.67	65.39	5.52	2.63	129
	Fiber	22						
	Filler	8						
F	Resin	70	30.60	3.86	62.96	5.65	2.09	131
	Fiber	20						
	Filler	10						

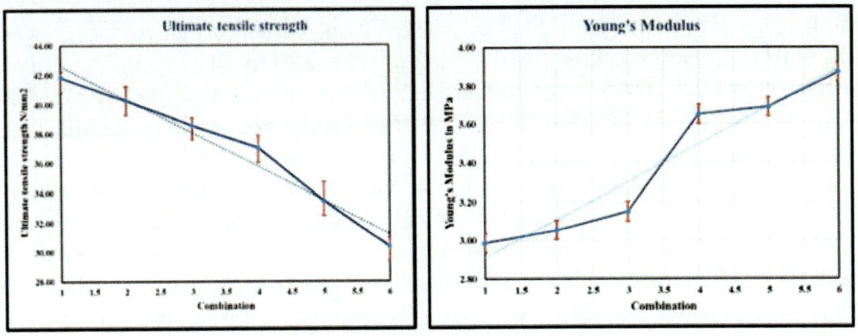

(a) (b)

Fig. 3 Variation of tensile strength and modulus of the specimen

2% increase in the filler material. Similarly, the tensile strength is reduced by 7.95, 13.6, and 22.01% for samples 'C', 'D' and 'E', respectively. The least tensile strength of 30.60 MPa was attained for sample 'F' (i.e.,70% of resin, 20% of fiber with 10% of filler material), which is 26.72% lesser than sample 'A'.

The increase in the proportion of filler material increases the stiffness of the composite material [16]. Thus, Young's modulus of a composite material increases as the filler proportion increases, as shown in Fig. 3b. The Young's modulus of 2.98 GPa was obtained for sample 'A' and keeps on increasing gradually with a gradual increase in the filler material. The maximum Young's modulus of 3.86 GPa was achieved for sample 'F', which is 29.5% greater than sample 'A'. Figure 4 shows the variation of flexural strength and flexural modulus of the specimens.

Figure 4 clearly shows the effect of the percentage of filler material on the three-point bending strength and modulus of the composite. The maximum three-point bending strength of 84.56 MPa was achieved for sample 'A'. The bending ability of the composite material decreases gradually with the increase in the percentage

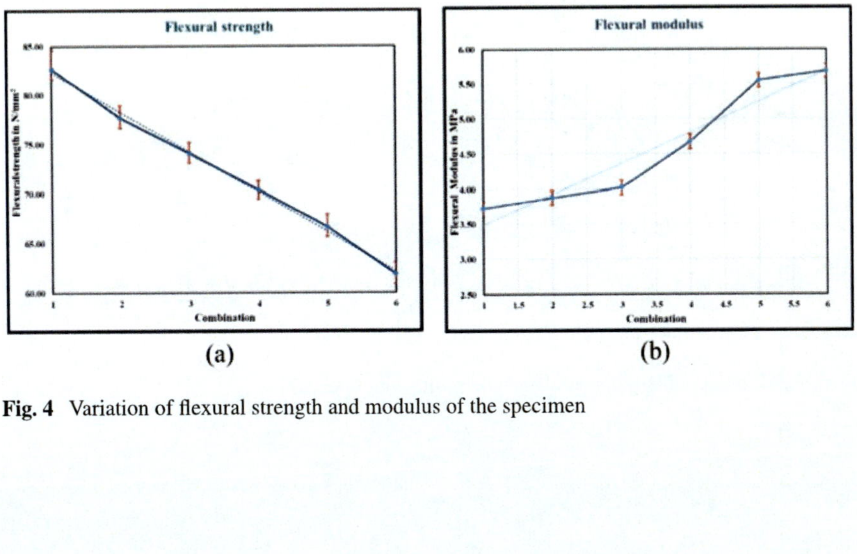

(a) (b)

Fig. 4 Variation of flexural strength and modulus of the specimen of filler material, as shown in Fig. 4a. The bending strength of the composite was reduced by 7.02% for sample 'B' (i.e.,70% of resin, 28% of fiber with 2% of filler material). Similarly, the bending strength reduced is by 11.28, 17.8, and 22.67% for samples 'C', 'D', and 'E' respectively. The least bending strength of 62.96 MPa was obtained for sample 'F', which is 25.54% lesser than the base material.

It can be seen that the stiffness of the composite increases as the filler proportion increases. Thus, the flexural modulus (bending modulus) of the composite improves with an increase in the percentage of filler material, as shown in Fig. 4b. The flexural modulus of 2.72 GPa was obtained for sample 'A' and its value increases gradually with a gradual increase in the filler material. The maximum flexural modulus of 5.65 GPa was achieved for sample 'F', which is 53.11% greater than sample 'A'.

It is evident from the results that the tensile and flexural strengths of the Pistachio shell filled banana fiber reinforced composite reduces with an increase in the percentage of filler material. The increase in filler content may lead to weaker bonding between fibers and the matrix material. In addition, at a high percentage of filler, non-uniform distribution of filler can cause agglomeration and stress concentration zones to appear. Thus, it reduces the tensile and flexural strength of the composite material. Figure 5a shows the SEM image of sample 'A' and figure (b) shows the SEM image of sample 'F' where the agglomeration of filler material can be observed. Figure 6 shows the variation in hardness and impact strength of the specimens.

From Fig. 6a, the hardness of the samples gradually increases as the percentage of filler increases. Sample 'F' exhibits a maximum hardness, which is 14.91% greater than sample 'A'. Increasing the filler concentration increases the fragility of the composite. And also, the increase in the concentration of filler material reduces the ability of the matrix to absorb energy and thereby reducing the toughness [17], thus reducing the impact strength from 6.41 to 2.09 kJ/m^2 as shown in Fig. 6b.

The reduction in the elongation of the test specimen containing filler material at the point of failure can be explained by the fact that the elastic properties of the composites depend on the polymer matrix exhibiting brittle behavior in the presence

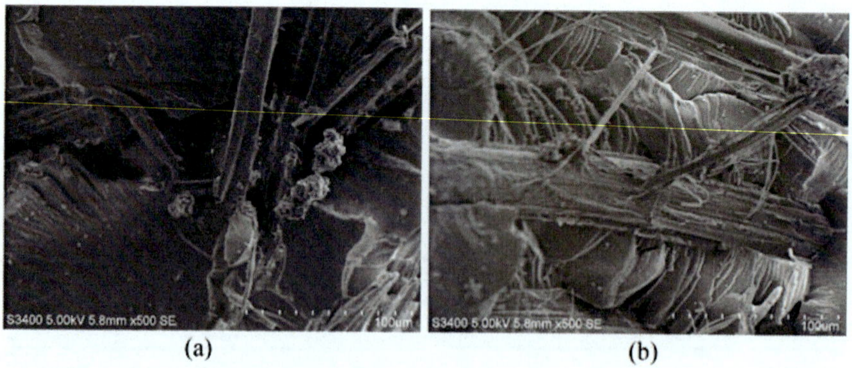

(a) (b)

Fig. 5 SEM images of base material and composite with filler material

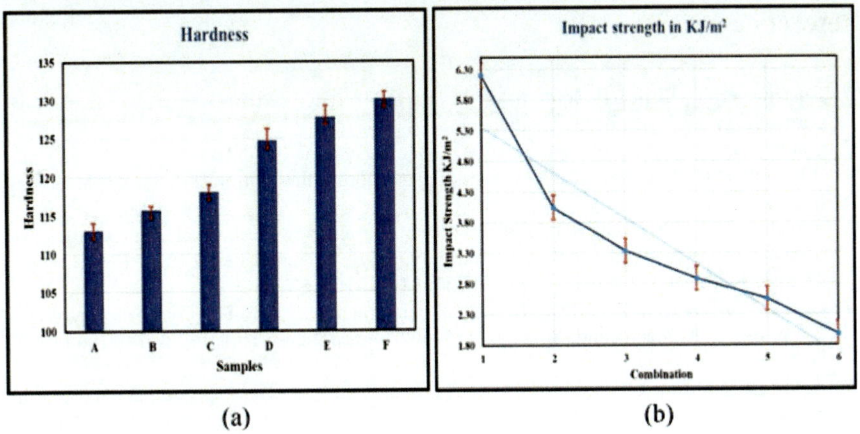

Fig. 6 Variation of the hardness of the specimen

of fillers. This is because the existence of filler material restricts the mobility of the polymer. Higher the filler content in the composites, higher will be the brittleness in the composites [18].

4 Conclusion

The experimental investigations on the effect of Pistachio shell as filler material on banana fiber reinforced composite clearly show that the mechanical performance of the composite varies with respect to the percentage of filler material. In general, the increase in the percentage of filler material decreases the tensile and flexural strength of the composite. The tensile and flexural strength of the composite is almost reduced by 26.72 and 25.54%, respectively, lesser than the base material. The maximum Young's and flexural modulus of 3.86 and 5.65 GPa, respectively, was achieved for sample 'F', which is 29.5 and 53.11%, respectively, greater than sample 'A'. At the same time, the stiffness of the composite increases as the filler proportion increases. Thus, increases the Youngs and flexural modulus of the composite. The intensification in the brittle behavior of the composite indicates an increase in the hardness of the composite. And also, an increase in the concentration of filler material reduces the capacity of the matrix to absorb energy and thus reduces the toughness of the material; therefore, the impact strength was reduced from 6.41 to 2.09 kJ/m^2. Thus, the pistachio shell as filler material on banana fiber reinforced composite influences the mechanical behavior of the composite material.

References

1. Rustemeyer T, Elsner P, John S-M, Maibach HI (eds) (2012) Kanerva's occupational dermatology. Springer, Berlin Heidelberg
2. Ozsoy I, Demirkol A, Mimaroglu A, Unal H, Demir Z (2015) The influence of micro-and nano-filler content on the mechanical properties of epoxy composites. Strojniski Vestnik/J Mech Eng 61(10)
3. Zaman HU, Hun PD, Khan RH, Yoon KB (2012) Morphology, mechanical, and crystallization behaviors of micro- and nano-ZnO filled polypropylene composites. J Reinf Plast Compos 31(5):323–329. https://doi.org/10.1177/0731684411436126
4. Gao Y, Liu L, Zhang Z (2009) Mechanical performance of nano-$CaCO_3$ filled polystyrene composites. Acta Mech Solida Sin 22(6):555–562. https://doi.org/10.1016/S0894-9166(09)60386-4
5. Agubra VA, Owuor PS, Hosur MV (2013) Influence of nanoclay dispersion methods on the mechanical behaviour of e-glass/epoxy nanocomposites. Nanomaterials 3(3):550–563. https://doi.org/10.3390/nano3030550
6. Yasmin A, Daniel IM (2004) Mechanical and thermal properties of graphite platelet/epoxy composites. Polymer 45(24):8211–8219. https://doi.org/10.1016/j.polymer.2004.09.054
7. Sayer M (2014) Elastic properties and buckling load evaluation of ceramic particles filled glass/epoxy composites. Compos B Eng 59:12–20. https://doi.org/10.1016/j.composites.2013.11.016
8. Asi O (2009) Mechanical properties of glass-fiber reinforced epoxy composites filled with Al_2O_3 particles. J Reinf Plast Compos 28(23):2861–2867. https://doi.org/10.1177/0731684408093975
9. Yang HS, Kim HJ, Son J, Park HJ, Lee BJ, Hwang TS (2004) Rice-husk flour filled polypropylene composites; mechanical and morphological study. Compos Struct 63(3–4):305–312. https://doi.org/10.1016/S0263-8223(03)00179-X
10. Uma Devi L, Bhagawan SS, Thomas S (1997) Mechanical properties of pineapple leaf fiber-reinforced polyester composites. J Appl Poly Sci 1739–1748
11. Kannan G, Thangaraju R, Kayaroganam P, Davim JP (2022) The combined effect of banana fiber and fly ash reinforcements on the mechanical behavior of polyester composites. J Nat Fibers 19(15):11384–11403
12. Salasinska K, Ryszkowska J (2015) The effect of filler chemical constitution and morphological properties on the mechanical properties of natural fiber composites. Compos Interfaces 22(1):39–50
13. Rami N, Meghraoui H, Ziraoui R, El khoukhi T, Mouhib M, El Harfi A (2010) Influence of gamma irradiation on the chemical and physical properties of DGEDDS/PDA and DGEDDS / MDA epoxy resins. J Mater Environ Sci 1(4):277–288
14. Bhattacharya A, Misra BN (2004) Grafting: a versatile means to modify polymers: techniques, factors and applications. Prog Polym Sci 29(8):767–814
15. Form and Style for ASTM Standards, © By American Society for Testing and Materials edition approved in 2010 as D638–10. https://doi.org/10.1520/D0638-14
16. Cho D, Lee J (2008) Effect of quasi-carbonization processing parameters on the mechanical properties of quasi-carbon/phenolic composites. J Appl Polym Sci 107(5):3350–3357
17. Sathees Kumar S, Kanagaraj G (2016) Investigation of characterization and mechanical performances of Al_2O_3 and SiC reinforced PA6 hybrid composites. J Inorgan Organometal Polym Mater 26:788–798
18. Haque MM, Hasan M (2021) TiO_2 and zeolite nanopowder enhanced mechanical properties of hybrid polymer composites. J Thermoplast Compos Mater 34(3):382–395

Author Index

V. Bindhu et al. (eds.), *Proceedings of Sixth International Conference on Inventive Material Science Applications*, Advances in Sustainability Science and Technology,
https://doi.org/10.1007/978-981-99-4189-6